# FORTSCHRITTE DER PHYSIK

HERAUSGEGEBEN IM AUFTRAGE DER PHYSIKALISCHEN GESELLSCHAFT
DER
DEUTSCHEN DEMOKRATISCHEN REPUBLIK
VON F. KASCHLUHN, A. LÖSCHE, R. RITSCHL UND R. ROMPE

HEFT 6 · 1982 · BAND 30

AKADEMIE-VERLAG · BERLIN

ISSN 0015-8208 Fortschr. Phys., Berlin 30 (1982) 6, 311–349 EVP 10,– M 31728

BEZUGSMÖGLICHKEITEN

Bestellungen sind zu richten

- **in der DDR** an eine Buchhandlung
  oder an den AKADEMIE-VERLAG,
  DDR-1086 Berlin, Leipziger Straße 3—4
- **im sozialistischen Ausland** an eine Buchhandlung für fremdsprachige Literatur oder an den zuständigen Postzeitungsvertrieb
- **in der BRD und Berlin (West)** an eine Buchhandlung oder an die Auslieferungsstelle
  KUNST UND WISSEN, Erich Bieber OHG,
  D-7000 Stuttgart 1, Wilhelmstraße 4—6
- **in den übrigen westeuropäischen Ländern** an eine Buchhandlung oder an die Auslieferungsstelle
  KUNST UND WISSEN, Erich Bieber GmbH,
  CH-8008 Zürich/Schweiz, Dufourstraße 51
- **im übrigen Ausland** an den Internationalen Buch- und Zeitschriftenhandel; den Buchexport, Volkseigener Außenhandelsbetrieb der Deutschen Demokratischen Republik,
  DDR-7010 Leipzig, Postfach 160; oder
  an den AKADEMIE-VERLAG,
  DDR-1086 Berlin, Leipziger Straße 3—4

Zeitschrift „Fortschritte der Physik"

Herausgeber: Prof. Dr. Frank Kaschluhn, Prof. Dr. Artur Lösche, Prof. Dr. Rudolf Ritschl, Prof. Dr. Robert Rompe, im Auftrag der Physikalischen Gesellschaft der Deutschen Demokratischen Republik.
Verlag: Akademie-Verlag, DDR-1086 Berlin, Leipziger Straße 3—4; Fernruf: 2236221 und 2236229; Telex-Nr.: 114420; Bank: Staatsbank der DDR, Berlin, Konto-Nr.: 6836-26-20712.
Chefredakteur: Dr. Lutz Rothkirch.
Anschrift der Redaktion: Sektion Physik der Humboldt-Universität zu Berlin, DDR-1040 Berlin, Hessische Straße 2.
Veröffentlicht unter der Lizenznummer 1324 des Presseamtes beim Vorsitzenden des Ministerrates der Deutschen Demokratischen Republik.
Gesamtherstellung: VEB Druckhaus „Maxim Gorki", DDR-7400 Altenburg, Carl-von-Ossietzky-Straße 30/31.
Erscheinungsweise: Die Zeitschrift „Fortschritte der Physik" erscheint monatlich: Die 12 Hefte eines Jahres bilden einen Band. Bezugspreis je Band 180,— M zuzüglich Versandspesen (Preis für die DDR: 120,— M). Preis je Heft 15,— M (Preis für die DDR: 10,— M).
Bestellnummer dieses Heftes: 1027/30/6.

AN (EDV) 57618

Fortschritte der Physik **30**, 311–349 (1982)

# The Investigation of Fission Isomers by Photonuclear Reactions

U. KNEISSL, W. GÜNTHER[1]), K. HUBER, H. KRIEGER[2])

*Institut für Kernphysik, Strahlenzentrum der Justus-Liebig-Universität Gießen, D-6300 Gießen, BRD*

**Contents**



---

[1]) Present address: Fa. Heyde u. Brandt, Bad Nauheim
[2]) Present address: Katharinenhospital Stuttgart

**Abstract**

The experimental techniques and possibilities for the investigation of fission isomers by photonuclear reactions are described. The available experimental data from isomeric photofission experiments are summarized and compared with results from particle induced reactions. Two types of experiments, taking advantage of the low angular momentum transfer in photonuclear reactions, are emphasized:

— The search for a γ-decay of shape isomers
— The measurement of isomeric cross section ratios for the double isomers in odd Pu-isotopes

Attempts of spin assignments for shape isomers are discussed and compared with both the available spectroscopic data and single particle calculations of excitations in the second well of the potential energy surface.

## A. Introduction

The concept of the double humped fission barrier of isotopes in the actinide region is well established both experimentally and theoretically. The discovery of the phenomenon was initiated by the first observation of a fission isomer, a spontaneously fissioning metastable state of an abnormal short half-life, by Polikanov et al. [1] in 1962. The angular momentum dependence of the production cross sections for these isomeric states together with their excitation energies (2—4 MeV) showed that fission isomers could not be interpreted as high spin states with a strongly hindered γ-decay due to spin selection rules.

The observed fission isomers could be explained within Strutinsky's hybrid model [2]. The main idea of this model is the superposition of deformation dependent shell corrections, calculated within a microscopic model, to the smooth liquid drop potential. This leads to a second minimum in the potential energy surface, as shown schematically in Fig. 1. In this picture the fission isomers are interpreted as states in the second well, as shape isomers with nearly twice the groundstate deformation. Their half-life is determined by the heights and widths of the double humped fission barrier.

In the past about 50 fission isomers have been observed in the actinide region ($92 \leqq Z \leqq 97$). The upper limit of this "island of isomerism" is caused by the decrease of the outer barrier for higher atomic numbers $Z$, which is mainly due to the inward shift of the liquid drop model fission barrier with increasing $Z$ [3]. The abrupt failure of isomers for lighter actinides $Z < 92$ hints that there is an increased penetrability of the inner barrier as $Z$ decreases.

The first experimental proof of the large deformation of fission isomers came from the experiment of H. J. Specht et al. [4] in 1972 who identified by conversion electron spectroscopy a rotational band built on the $^{240}$Pu fission isomer. The measured moment of inertia was about a factor of two larger than that of the ground-state. A more model independent measurement of the deformation represents the determination of the quadrupole moment of shape isomeric states as performed by the Heidelberg group [5]. They succeeded in measuring lifetimes ($B(E2)$-values) of rotational transitions in the second well using the charged plunger technique [6, 7]. The resulting quadrupole moments (32—36 b) correspond to a ratio of 2:1 for the axis of an ellipsoide. An independent confirmation of these results was given by a very recent measurement of the optical isomer shift of atomic transitions in the case of $^{240m}$Am [8].

The interpretation of the fission isomers as shape isomers can be regarded to be well established in view of existing experimental data. A systematics of half-lives has been measured and can be qualitatively explained by the systematic of the double humped fission barrier [3]. The interest in current experiments is focused mainly on the investigations of the spectroscopic properties of the highly deformed nuclear states [7]. The study of spins and moments of these isomeric states or of possible single particle and collective excitations in the second well, such as rotational or vibrational states, provides important information concerning the nucleus at extreme deformations. This spectroscopy within the second minimum is a very difficult task and requires new, sophisticated experimental techniques. The present results and experimental methods are described in the review article of METAG et al. [7].

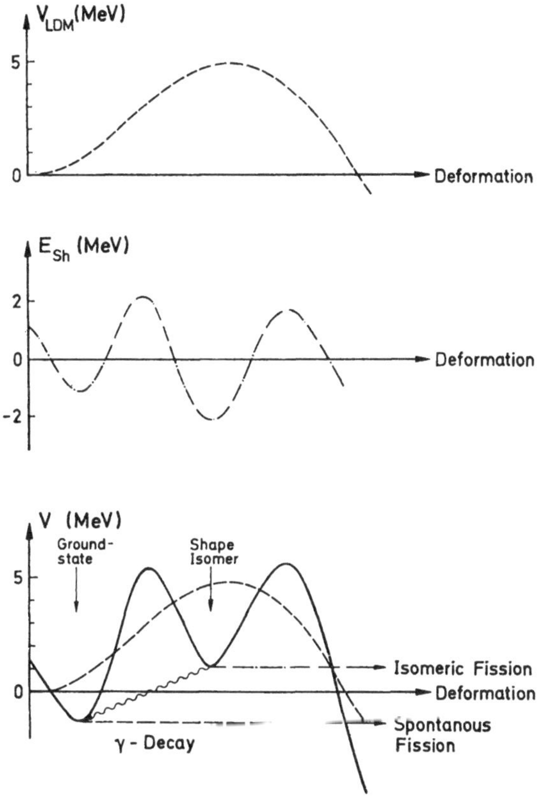


Fig. 1. The double humped fission barrier $V$ generated by the superposition of the deformation dependent shell correction $E_{sh}$ to the liquid drop barrier $V_{LDM}$.

Another open field represents the investigation of the decay properties of fission isomers in light actinides. As mentioned above the failure of fission isomers for $Z < 92$ can be explained by the lowered inner barrier. This should favour a γ-decay of the shape isomers back to the first minimum. Up to now only one γ-decay like this (for $^{238m}U$) [9] has been reported. However, two experiments [10, 11] on $^{236m}U$ failed to detect such a strong direct γ-decay branch. This different behaviour is difficult to understand in view of nearly equal barrier parameters and similar level schemes in the first well for $^{238}U$ and $^{236}U$.

Photonuclear reactions seem to be favourable for the investigation of fission isomers. Due to the missing Coulomb barrier compound states of low excitation energy are easily populated. The number of open channels is reduced compared with particle induced reactions. The main advantage, however, is the spin selectivity of the excitation. The giant electric dipole resonance dominates the photoabsorption cross section in the most important energy range. The low transfer of angular momentum in a dipole excitation leads to the favoured population of low spin states. This is of advantage for investigations of fission isomers.

Unfortunately there exist severe experimental difficulties when using photonuclear reactions. First of all there is a lack of intense monochromatic photon sources. The mostly used continuous bremsstrahlung sources do not permit to measure excitation functions or other energy dependences in an unambiguous way (see section C). Additionally the small photonuclear cross sections require high current electron accelerators to produce intense bremsstrahlung sources. The small duty cycle of such machines ($\leq 10^{-3}$) drastically increases the background problems at such facilities. Furthermore the low momentum of the photons compared with charged particles does not allow to apply recoil methods, mainly used for the spectroscopy within the second well. In spite of these mainly experimental drawbacks the spin selectivity of the photonuclear excitation offers important benefit in specific cases. It is well known that in the barrier region photofission cross section and angular distribution measurements are a powerful tool to get information on the barrier parameters and quantum numbers [*12*]. For the investigation of fission isomers two applications are of special interest.

(1) The search for a γ-decay of fission isomers for light actinides. In order to detect a γ-decay of a shape isomer delayed γ-coincidence spectra have to be measured. For such an experiment a photonuclear reaction seems to be of advantage, since the low transfer of angular momentum strongly handicaps the excitation of high spin isomers in nonfissioning residual nuclei and fission fragments, which could complicate the interpretation of the experimental data.

(2) The investigations of double isomers in even-odd isotopes. In even-odd isotopes, besides the isomeric ground-states in the second well, additional isomers are known, which are interpreted as single particle excitations. The measurement of the relative populations of these double isomers in reactions of different transfer of angular momentum yields information on the spins of the levels involved. The photonuclear reaction with its low spin transfer represents a significant extension of the range of transferred angular momentum covered by particle induced reactions. In combination with particle induced data this leads to more reliable spin assignments (see section E).

Recently several review articles [*7, 13–15*] have been published dealing with the interesting phenomenon of shape isomerism and the spectroscopic properties of these highly deformed nuclear states. The aim of the present work is to summarize the results and contributions given to this field by photonuclear experiments and to show the possibilities of this reaction type. This seems to be of interest, since in the last years new detection techniques [*16, 17*] and more versatile electron accelerator facilities got available, leading to a considerable extension of the investigated half-life time range and allowing more detailed photonuclear experiments. Since older photonuclear results on longlived isomers are described in detail by GANGRSKY et al. [*18*], in the present article the more recent experiments [*10, 19–23*] are emphasized, performed mainly at the 65 MeV Giessen electron linac.

After a brief review of the experimental techniques (section B) the available isomeric fission cross section data ((γ, n)-, (γ, 2n)- and (γ, γ')reactions) are discussed in section C. Experiments to detect a γ-decay of shape isomers are described in chapter D. The last section deals with the results of spin assignments from the analysis of isomeric cross section ratios for double isomers.

## B. Experimental Techniques

In particle induced reactions the velocity of the recoil nuclei is typically in the order of $v_R \approx 0.1-1$ mm/ns depending on the reaction and the kinematics. Therefore, using thin targets, recoil techniques often have been applied for the investigation of fission isomers. Details of these experimental techniques have been reviewed for example in ref. [*15*]. The recoil distance method has been further developed in the projection method for the subnanosecond time range [*24*]. The spectroscopy in the second well, using the charge plunger and recoil shadow method [*7*], is also taking advantage of the recoiling nucleus. All these techniques are not applicable in photonuclear reactions since the low photon momentum causes only low recoil energies in the order of keV, preventing the recoil from emerging from the target. At tandem accelerators the associated particle method has successfully been employed (e.g. in $(d, p)$-reactions), where delayed coincidences between the associated particle and the fission fragments have to be measured. This coincidence technique is not feasible in photonuclear reactions since up to now all intense electron accelerators are pulsed machines of low duty cycle ($\leq 10^{-3}$), where coincidence experiments suffer from the overwhelming rate of accidental coincidences.
For all these reasons the pulsed beam technique seems to be the only practicable method for photonuclear studies of fission isomers. Such experiments require pulsed electron accelerators with variable pulse widths in the range of ns to some µs at high repetition rates, with variable energy and high intensity. Furthermore fast fragment detectors are a basic supposition of such investigations.
The time zero reference, needed in pulsed beam experiments, can be derived from

a) machine synchronisation pulses (gun trigger etc.),
b) fast scintillation counters detecting the prompt $\gamma$-flash,
c) a ferrite monitor, detecting the electron beam pulses.

The time measuring devices (such as time to pulse height converters) can be calibrated absolutely with high accuracy. Therefore the absolute time scale contains nearly no systematic uncertainty as in recoil distance experiments.

### B.1. Photon Sources

The cross sections for the population of fission isomers in photonuclear reactions are very low, typically in the order of 10—100 µb. Therefore the intensity of all known sources of monoenergetic photons (s. review of Beil and Bergère [*25*]) are by far not suf-

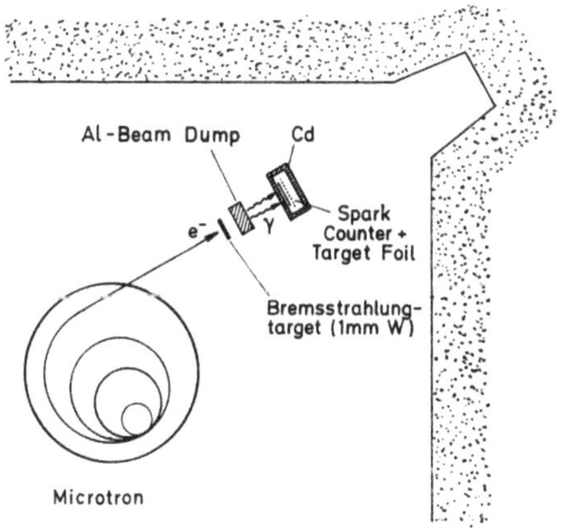


Fig. 2. Experimental set-up used by Gangrsky et al. [18] at the Moscow microtron.

ficient for isomeric fission experiments. The use of intense bremsstrahlung beams produced by intense electron accelerators is the only possibility. The first photonuclear pulsed beam experiments to study fission isomers have been performed at the 17 orbit microtron in Moscow by Gangrsky et al. [*18*]. Their set up is schematically shown in fig. 2. A 1 mm tungsten target was used for the bremsstrahlung production. The electrons were stopped in an aluminium absorber. This simple beam stop was possible because of the low beam intensity and an energy (< 17 MeV) which is below the (γ, n)-threshold of aluminium. With this microtron and the used spark counters (s. sections B.2.1.) only long-lived μs-ms isomers could be investigated in (γ, γ′) and (γ, n)-reactions

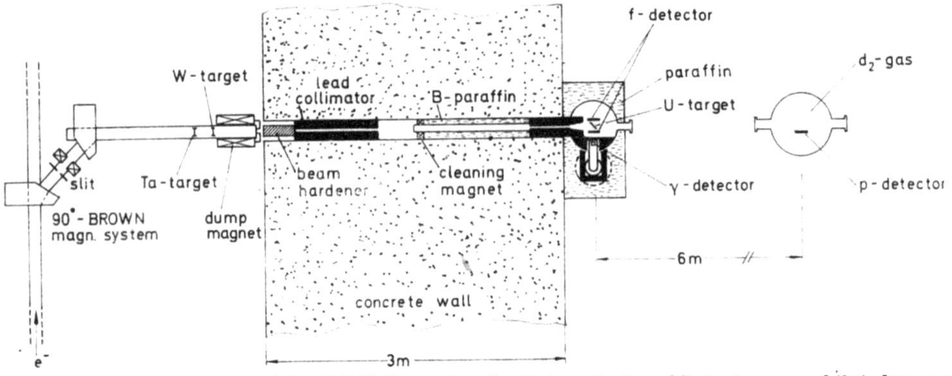


**Fig. 3.** Experimental set-up used at the 65 MeV Giessen linac for the investigation of fission isomers and their decay properties [*10*].

More versatile, modern accelerators enable the study of shortlived isomers due to the possible shorter beam pulses in case of using fast fragment detectors. A typical arrangement as installed at the Giessen linac [*10*, *21*], is shown in fig. 3. The electron beam is analyzed in energy and focused on a *W*-bremsstrahlung target. The electrons passing the bremsstrahlung target are dumped by a sweeping magnet. The photon beam is well defined by a collimator- and beamhardener system in a 3 m concrete wall between the bremsstrahlung target and the experimental set up. This large distance and the good shielding is necessary to reduce the fast neutron background, when investigating short-lived fission isomers. Besides the fragment detectors a Ge(Li)-γ-spectrometer is shown in fig. 3 as used for searching a γ-decay of fission isomers. The bremsstrahlung spectrum can be measured simultaneously with a proton detector via the photodisintegration of the deuteron.

Table 1
Characteristics of electron accelerators used for photonuclear isomeric fission investigations

| Accelerator | Energy (MeV) | Pulse width | Rep. rate (Hz) | Used mean currents | Ref. |
|---|---|---|---|---|---|
| Microtron (Moscow) | 6–17 | 1–2 μs | 400–600 | 10–20 μA | [*18*] |
| Linac (Acad. Sci. USSR) | 10–80 | 1.3 μs | 50*) | 2–5 μA | [*23*] |
| Linac (Giessen) | 10–65 | 6–100 ns<br>0.1–5 μs | 1000<br>400–100 | 6–40 μA<br>≈ 200 μA | [*10*, *21*] |

*) used for the investigation of the 14 ms–$^{242}$Am-isomer.

In table 1 the beam characteristics of electron accelerators, used so far for photonuclear isomeric fission experiments, are summarized. The figures show that the high current Giessen linac with its variable pulse lengths is well suited for pulsed beam experiments.

### B.2. Fission Fragment Detectors

An inherent γ- and neutron background normally exists at intense bremsstrahlung facilities. The low duty cycle of the used conventional electron linacs ($\lesssim 10^{-3}$) increases dramatically the background problems at such accelerators. Therefore special properties are required for the fragment detectors in photofission experiments in particular in investigations of isomeric fission. The properties are as follows:

- insensitivity to γ- and neutron background
- high efficiency for fission fragments
- good time resolution
- good discrimination between fragments and α-particles from the target activity
- insignificant radiation damage caused by the high number of prompt fission fragments, by the α-activity of the target and by beam background
- large solid angle
- low production costs in view of a possible contamination

In the past, spark counters [*18, 23*], Si-surface detectors [*19, 20, 26*] and parallel plate avalanche detectors (PPAD) [*16, 21*] have been used. The properties and characteristics of these detector types are summarized in table 2 and discussed in the following chapter.

Table 2
Properties of fission fragment detectors

| Detector type | Spark-counter | Si-detector | Parallel plate detector |
|---|---|---|---|
| Rise-time | ≈ 50 ns | ≈ 5 ns | ≦ 1 ns |
| Recovery-time after the γ-flash | ≈ 8 – 20 μs | no | no |
| Efficiency for fragments | ≈ 0.3 | ≈ 1 | ≈ 1 |
| Background sensitivity | noticable | high | negligible |
| Usable solid angle | ≈ 4π | some percent of 4π | ≈ 4π |
| Radiation damage | no | noticable | no |
| Costs | inexpensive | expensive | inexpensive |
| References | [*18, 23*] | [*19, 20, 26*] | [*16, 17, 21, 22*] |

#### B.2.1. Spark counters

The soviet groups [*18, 23*] used multiwire spark counters for the detection of the fission fragments. The advantages of this detector, filled with 1 atm gas (98.5% He, 1.5% N), are the good efficiency (≈ 30%) for fragments, the good discrimination against α-particles and the large usable solid angle (≈ 4π). However, the achievable time resolution (≈ 100 ns) is quite modest. Furthermore these detectors show break downs during the intense γ-flashes. This causes recovery times in the order of 5 – 20 μs depending on the γ-intensity. Therefore GANGRSKY et al. [*18*] used this detector in a pulsed operation for the investigation of μs isomers in Pu-isotopes. Nevertheless the ratio of isomeric to prompt fission yields cannot be measured simultaneously with such devices, due to the insensitivity of the spark counters during the beam pulse. The prompt fission yield has to be determined in a separate run with reduced intensity and removed detectors. This leads to uncertainties in the absolute normalization of the beam intensities and the solid angle.

### B.2.2. Si-surface barrier detectors

Semiconductor detectors have been employed successfully for the first time at the Giessen bremsstrahlung facility [*19*]. Large area detectors (900 mm$^2$) were installed under 90° with respect to the $\gamma$-beam at a distance of 7 cm to the beam axis. However, a well collimated, electron free photon beam is a basic postulate for a reliable use. The detectors can be operated also during the beam pulse; this enables the simultaneous measurement of prompt and isomeric fission events. Due to the good energy resolution of the Si-detectors, there are no problems to discriminate $\alpha$-particles. A time resolution of about 5 ns can easily be achieved with conventional electronics. This enables the study of short-lived isomers. However, the detectors are quite sensitive to secondary electrons, accidentially produced on the target frame in the case of a not perfect beam or target alignment. Therefore, the detectors cannot be installed too near to the beam. Solid angles of only $\approx 200$ msr can be achieved. Furthermore the radiation damage is not negligible in the case of highly $\alpha$-active targets and long data collection times (some weeks). The radiation damage may cause tails in the prompt time distributions, which are not acceptable in isomeric fission work. Therefore Si-detectors can be operated only for a certain time, depending on the experimental conditions and requirements. In view of the high price of these detectors, they do not represent an optimal detector system for the experiments in discussion.

### B.2.3. Parallel plate avalanche detectors (PPAD)

The parallel plate avalanche detector is a proportional counter with two parallel, plane electrodes. As filling gas an organic vapour (some torr) is used. Voltages of 400—700 V are applied. Details of the construction and operation characteristics can be found elsewhere [*16*].

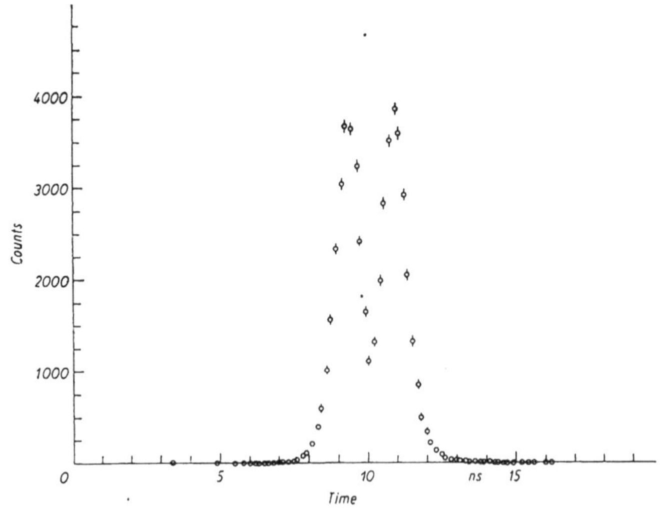


Fig. 4. Time distribution of fission fragments of a $^{252}$Cf-source placed between two PPAD's [16]. (Distance between source and detectors $d$ = 2.7 cm).

The PPAD's fulfil nearly all requirements for an ideal detector for isomeric photofission experiments as listed in the beginning of this chapter. The good time resolution (< 1 ns) is demonstrated in fig. 4, where the time distribution of fission fragments of a $^{252}$Cf-source measured with two PPAD's is shown. In fig. 5a PPAD-pulse height spectrum for α-particles and fission fragments from $^{252}$Cf is plotted. A good particle discrimination can be stated. The main advantage of the PPAD's, however, is the nearly complete insensitivity against γ- and n-background.

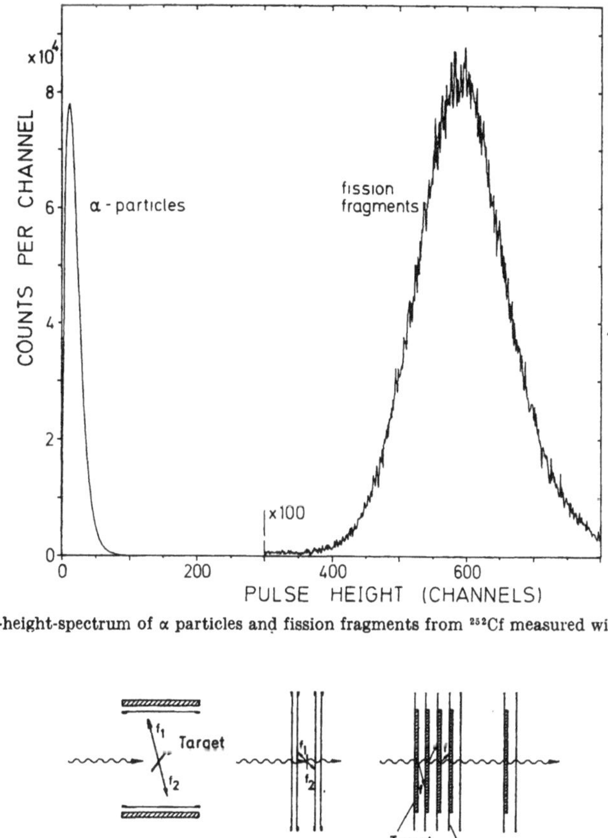


Fig. 5. Pulse-height-spectrum of α particles and fission fragments from $^{252}$Cf measured with a PPAD [16].

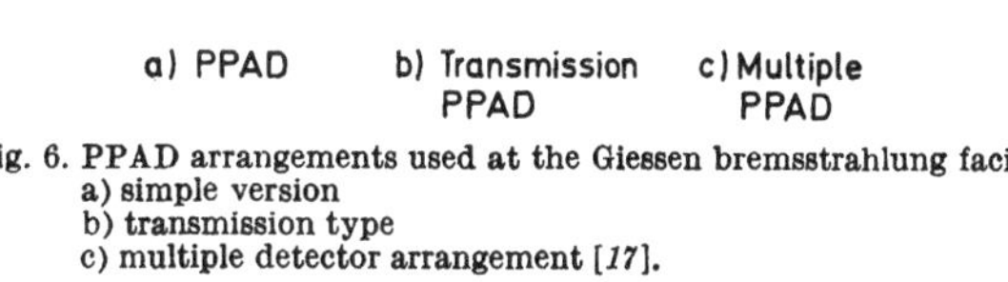


Fig. 6. PPAD arrangements used at the Giessen bremsstrahlung facility
a) simple version
b) transmission type
c) multiple detector arrangement [17].

Encouraged by the successful use of PPAD's in particle induced isomeric fission experiments [27] and in heavy ion physics [28] several arrangements of PPAD's have been constructed and employed at the Giessen bremsstrahlung facility as shown in fig. 6. As simpliest set-up a detector pair was installed under 90° to the beam. The detector surface was 25 cm², the distance to the beam axis was ≈ 3 cm, resulting in a considerably increased solid angle, when compared with Si-detectors. A transmission type (7 × 7 cm²) with thin entrance and exit foils covers a solid angle of ≈ 4π. This enables the investigation

of small target material quantities, which is important in the case of enriched isotopes. A special detector, a multiple parallel plate detector has been developed for the measurement of the very low cross sections [17]. It consists of 30 single detectors, where the fissible target material is electrolytically deposited on one detector foil (⌀ 65 mm). The target thickness amounts to about 2.5 mg/cm$^2$, thus a total layer of 75 mg/cm$^2$ can be used in the beam, though the fragment range in uranium for example is only about 8 mg/cm$^2$.
All PPAD's are used in connection with a fast preamplifier in order to take advantage of the good intrinsic time resolution of the detector. These home made preamplifiers [17] have pulse rise times of $\approx 0.2$ ns and a gain of $\approx 200$. Therefore, a time resolution of $< 1$ ns can be achieved by using conventional electronics.
Since PPAD's are energy loss detectors, back scattered $\alpha$-particles with a long path through the detection volume cause pulses which may reach the pulse heights of fission fragments. In order to discriminate against this $\alpha$-background in the case of targets of high $\alpha$-activity an additional coincidence condition between two fragment detectors can be required [21]. A possible contribution of spontaneous fission of the target nucleus can be measured simultaneously between the beam pulses and can be used for correction of the delayed fission counting rates.
The expected isomeric yield is about 5 order of magnitude lower than that of prompt fission. This low yield clearly demonstrates the extreme requirements for a trouble-free operation of the fragment detectors and the whole set up. In order to get reliable results the quality of the timing as well as the background level from neutron induced fission have to be tested seriously. As "dummy targets" $^{232}$Th and $^{235}$U can be used to check the fast and thermal neutron background since no isomers are populated when bombarding these isotopes with bremsstrahlung [21].

### B.3. Evaluation and Data Reduction

A typical example of a measured time distribution of the fission fragments is shown in fig. 7 ($^{238}$U($\gamma$, 2n) $^{236m}$U). The number of fission fragments is plotted in a logarithmic scale versus the time (running from the right to the left side). The peak corresponds to prompt fission fragments and shows the beam pulse time distribution. Delayed fission events can be seen in this log.-scale as a linear decrease corresponding to the life-time of the isomer.

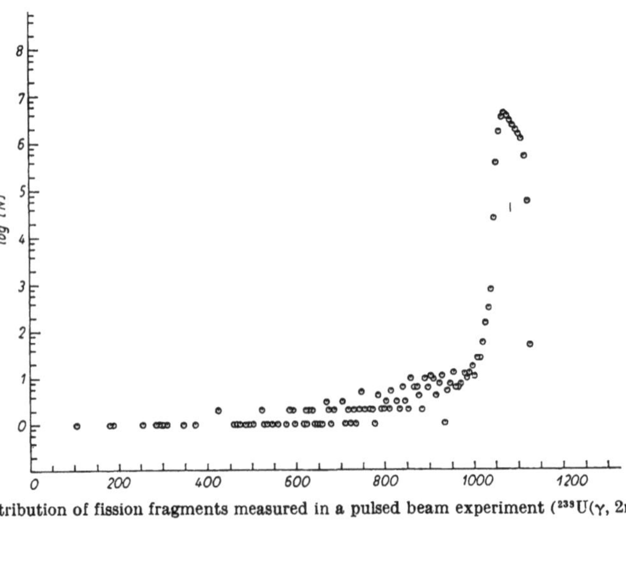


Fig. 7. Time distribution of fission fragments measured in a pulsed beam experiment ($^{238}$U($\gamma$, 2nf)) (Ref. [16]).

In order to evaluate the half-life $t_{1/2}$ of the fission isomer from the measured time distribution usually several time channels have to be added up due to the low channel contents. This addition procedure, however, represents some arbitrary choice and affects the overall time resolution. Therefore the use of a $^2$log-time scale, as introduced by K. HUBER [29], seems to be advantageous. In this $^2$log-time scale the successive time intervals are just doubled. An exponential decrease is transformed to a function as shown in fig. 8. According to the doubling of the measuring time intervals the intensities increase by $2^n$ for $t \ll t_{1/2}$, reach their maximum at $t = t_{1/2}$ and decrease rapidly by $2^{-2n}$ for $t \gg t_{1/2}$. The maximum lying at the half-life $t_{1/2}$ corresponds to 1/4 of the total activity. The amplitude and the halflife can be determined by a least squares fit. This method of

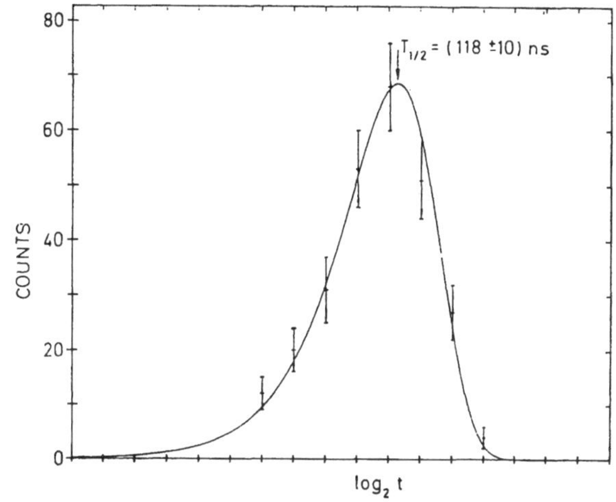


Fig. 8. Time distribution of the delayed fission fragments plotted in a $^2$log-time scale. The full line represents a least squares fit [16].

data reduction is even very advantageous in the analysis of decay curves with several time components. Furthermore in 2-parameter experiments (for example energy-time matrices s. section D.4.) it leads to a considerable data reduction and enables high energy resolution measurements over a large energy and time range with constant relative accuracy of the time determination.
Technically the transformation to a $^2$log-time scale is easily achieved by assorting in the computer the time information according to the highest bit.
The ratio of isomeric to prompt fission yield can be evaluated from the measured prompt fission events and the extrapolated delayed fission time distribution. At this procedure the form of the prompt time distribution, as measured simultaneously, has to be taken into account.

## C. Photonuclear Isomeric Fission Cross Sections

### C.1. Analysis

#### C.1.1. Bremsstrahlung enfolding techniques

All photonuclear studies of fission isomers have been performed so far with bremsstrahlung, because of the low intensity of all available sources of monoenergetic photons. Investigating a photonuclear reaction of a cross section $\sigma(E_\gamma)$ with bremsstrahlung one measures a reaction yield $Y(E_0)$ for an endpoint energy $E_0$ of the bremsstrahlung spectrum, as given by

$$Y(E_0) = \int_0^{E_0} \sigma(E_\gamma) \cdot N(E_\gamma, E_0)\, dE_\gamma \tag{1}$$

where $N(E_\gamma, E_0)$ represents the shape of the continuous bremsstrahlung spectrum. In order to extract the interesting cross section the endpoint energy has to be varied. The measured set of yields $Y_i(E_0^i)$ can be enfolded if the bremsstrahlung shape is known. The enfolding techniques, based on the inversed matrix method of Penfold and Leiss [*30*] have been improved (variable bin method [*31*], least structure technique [*32*]). For low energy subbarrier photofission, where an exponential increase of the cross section with the γ-energy is observed, Zhuchko et al. [*33*] developed a special procedure for the restoration of the photofission cross sections from measured yields.
However, in spite of these improved techniques, it is still very difficult to avoid spurious structures in the derived cross sections or to extract structures in an reliable way, except near to the threshold of a reaction (see ref. [*25*]).
Gangrsky et al. [*18*] and Kuznetsov et al. [*23*] measured for long-lived isomers the isomeric fission yield for various endpoint energies (9—13 MeV, 10—80 MeV, respectively). In spite of the low yields and the rather crude energy steps these authors succeeded in extracting directly a cross section curve by applying a conventional enfolding technique to the smoothed yield curve. Due to the necessary smoothing technique no fine structure could be derived but the general slope of the cross section. This bremsstrahlung enfolding technique is so far the only method to determine the energy dependence of photonuclear isomeric fission cross section in a model independent way.
Unfortunately the bremsstrahlung enfolding method requires very high-γ-fluxes due to the low cross sections and the necessary variation of the endpoint energy of the bremsstrahlung spectrum. The method could so far not be applied for the investigation of short-lived isomers, where the fast neutron background, produced on the bremsstrahlung target, has to be reduced by a good shielding and a well defined beam geometry. This leads necessarily to lower γ-intensities due to larger distances between γ-source and experimental set-up. Therefore, the available fluxes in such an arrangement are not high enough to make an endpoint energy variation feasible in reasonable running times.

### C.1.2. Statistical model calculations

In all cases, where only one measurement at fixed endpoint energy can be performed because of the reasons discussed above, one can derive an isomeric fission cross section using a statistical model calculation, if the yield ratio $Y_{iso}/Y_{prompt}$ has been measured. This ratio is given by

$$\frac{Y_{iso}}{Y_{pr.}} = \frac{\int_0^{E_0} N(E_\gamma, E_0)\, \sigma_{iso}(E_\gamma) \cdot dE_\gamma}{\int_0^{E_0} N(E_\gamma, E_0)\, \sigma_{\gamma,f}(E_\gamma) \cdot dE_\gamma} \tag{2}$$

where $N(E_\gamma, E_0)$ is the shape of the bremsstrahlung spectrum, $\sigma_{iso}(E_\gamma)$ is the isomeric fission cross section and $\sigma_{\gamma,f}(E_\gamma)$ the prompt fission cross section. $\sigma_{iso}$ can be written as

$$\sigma_{iso}(E_\gamma) = \sigma_{iso}^{max} \cdot f(E_\gamma) \tag{3}$$

where $f(E_\gamma)$ represents the energetic shape of the cross section and $\sigma_{iso}^{max}$ gives the absolute scale.

Using known experimental prompt photofission cross sections, the shape function $f(E_\gamma)$ calculated within an evaporation model and a Schiff-bremsstrahlung spectrum [34], one can now evaluate isomeric photofission cross sections from the measured yield ratio according to (equ. 2).

The shape function $f(E_\gamma)$ can be calculated, as proposed by VANDENBOSCH [35] for (p, 2n) reactions using a modified Jackson evaporation model. Within this model it is assumed that the neutron evaporation spectra can be described by a Maxwell distribution with a constant temperature $T$, corresponding to a simple exponential level density formula. In this approximation the occuring integrals over level densities can be solved analytically resulting in Pearson's incomplete $\Gamma$ functions.

$$I(\Delta_j, 2_j - 3) \qquad (j = \text{Number of evaporated neutrons}). \tag{4}$$

The arguments

$$\Delta_j = \left(E - \sum_{K=1}^{j} B_K\right) \Big/ T \tag{5}$$

contain neutron binding energies or barrier parameters $B_K$. In the case of a $(\gamma, 2n)$ reaction one gets

$$f(E_\gamma) \sim \sigma(\gamma, 2n) \frac{\varrho_{II}}{\varrho_I} \frac{I(\Delta_2^i, 3) - I(\Delta_3, 3)}{I(\Delta_2, 1) - I(\Delta_3, 3)} \tag{6}$$

where in $\Delta_2^i = (E - E_{II} - B_1 - B_2)/T$ the excitation energy $E$ has to be reduced by the energy $E_{II}$ of the second well. $\varrho_{II}/\varrho_I$ is the ratio of the level densities in the two minima of the potential energy surface ($\varrho_{II}/\varrho_I$ is constant for a simple $\varrho(E) \sim e^{E/T}$ level density formula). The Pearson functions $I(\Delta_3, 3)$ take into account the competition of the $(\gamma, 3n)$ channel above the corresponding threshold. This model can be applied also to $(\gamma, n)$ reactions in an analogous way

$$\sigma_{iso}(E_\gamma) \approx \sigma(\gamma, n) \cdot \frac{\varrho_{II}}{\varrho_I} \frac{I(\Delta_1^i, 3) - I(\Delta_2, 3)}{I(\Delta_1, 1) - I(\Delta_2, 3)}. \tag{7}$$

The photoneutron cross sections $\sigma(\gamma, 2n)$ and $\sigma(\gamma, n)$ can be taken from experimental work [36] or be calculated using reasonable $\Gamma_n/\Gamma_f$ values [12] and a simple statistical model [37].

Absolute prompt fission cross sections had been measured in the most important energy region of the giant dipole resonance (up to 18 MeV) by the Livermore [36] and Saclay [38] groups with monoenergetic photons for $^{232}$Th and U isotopes. For higher energies these data can be extrapolated using reasonable $\Gamma_n/\Gamma_f$ values [12, 36]. For the Pu isotopes such reliable monochromatic photon data are not available. However, one can take the shape of the photofission cross section as measured in older bremsstrahlung experiments [39] and normalize the scale using recent photoabsorption data [40] and $\Gamma_f/\Gamma_{tot}$ values from the systematics [12, 36]. For $^{241,243}$Am the $(\gamma, f)$ cross sections have been determined quite recently in a bremsstrahlung experiment up to 20 MeV [41].

The energy dependence of the isomeric fission cross section in $(\gamma, \gamma')$-reactions also can be calculated in a simple penetration model and using statistical model expressions (s. Ref. [18])

$$\frac{\sigma_{iso}(E_\gamma)}{\sigma_{\gamma,f}(E_\gamma)} = \frac{\Gamma_{2\gamma}}{\vec{\Gamma}_2} \tag{8}$$

where $\vec{\Gamma}_2$ represents the width for the penetration of the outer barrier and $\Gamma_2^\gamma$ is the $\gamma$-width in the second well (s. fig. 15). These widths are given in a statistical model description introducing the number of open channels $N_B$ and assuming $E1$-radiation

$$\vec{\Gamma}_2 = \frac{N_B}{2\pi \cdot \varrho_{II}(E_\gamma - E_{II})} \approx \frac{1}{2\pi \cdot \varrho_{II}(E_\gamma - E_{II})} \cdot \int\limits_0^{E_\gamma - B_f} \varrho_{II}(E_\gamma - B_f - \varepsilon)\, d\varepsilon$$

$$\Gamma_2^\gamma \approx \frac{1}{\varrho_{II}(E_\gamma - E_{II})} \cdot \int\limits_0^{E - E_{II}} \varrho_{II}(E_\gamma - E_{II} - \varepsilon)\, \varepsilon^3 \cdot d\varepsilon. \tag{9}$$

Calculations of equ. (9) [*42*] are quite insensitive to the used level density expression and yield in a nearly ratio

$$\frac{\sigma_{iso}}{\sigma_{\gamma,f}} \approx \text{const.}$$

in the energy range between 6 and 12 MeV. Thus the isomeric fission cross section increases like the prompt photofission cross sections.

## C.2. Results for Isomeric Fission Cross Sections

### C.2.1. (γ, n)-reactions

In fig. 9 results for $^{240}$Pu(γ, n) $^{239m_1}$Pu and $^{242}$Pu(γ, n) $^{241m_1+m_2}$Pu are shown as typical examples, derived from a measured yield ratio $Y_{iso}/Y_{prompt}$ using the statistical evaluation. The cross sections are centered at about 12 MeV, the maximum values amount to about 50 μb. The quoted errors are only statistical ones and do not contain the un-

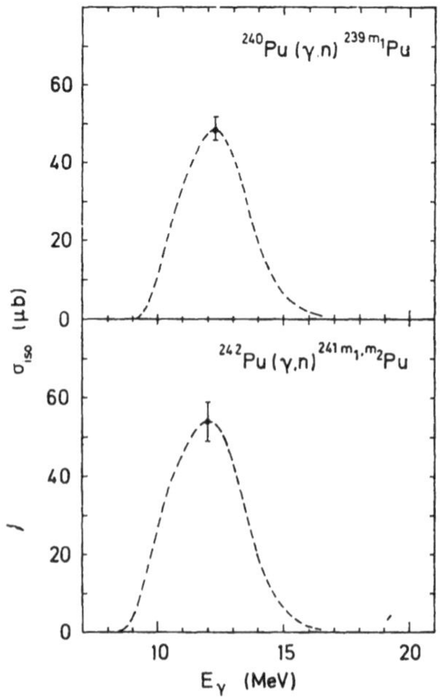


Fig. 9. Isomeric fission cross sections for the $^{240}$Pu(γ, n) $^{239m_1}$Pu and $^{242}$Pu(γ, n) $^{241m_1,m_2}$Pu reactions [*21*] (statistical model evaluation).

certainties of the prompt photofission cross sections (10–20%) used in the evaluation. In fig. 10 the very recent results for the long-lived Am-isomers are plotted ($^{240}$Am ($T_{1/2} = 0.9$ ms) and $^{242}$Am ($T_{1/2} = 14$ ms)) as reported by KUZNETSOV et al. [*23*]. In these experiments the yields have been measured as a function of the bremsstrahlung endpoint energy. Therefore the cross sections can be obtained by an enfolding technique. The cross sections are of the same order as in the Pu-case. Furthermore, the reasonable agreement of the shape with the results of the statistical model evaluations supports the reliability of these evaluation procedures.

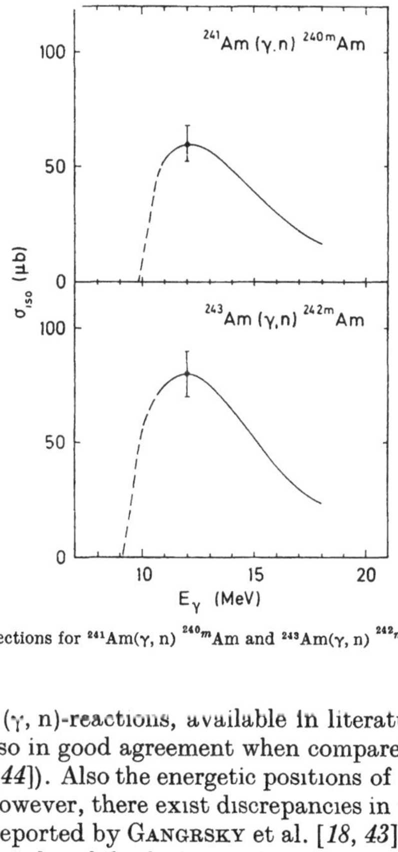


Fig. 10. Isomeric fission cross sections for $^{241}$Am(γ, n) $^{240m}$Am and $^{243}$Am(γ, n) $^{242m}$Am [*23*] (enfolding technique).

In table 2 the results for (γ, n)-reactions, available in literature, are summarized. The measured half-lives are also in good agreement when compared with data from particle induced reactions (s. ref. [*44*]). Also the energetic positions of the cross section maxima do not differ too much. However, there exist discrepancies in view of the absolute cross section scale. The values reported by GANGRSKY et al. [*18, 43*] are systematically higher by a factor of 2–3, compared with both the results of KUZNETSOV et al. [*23*] and GÜNTHER et al. [*21*], respectively. Even higher discrepancies in the isomeric to prompt yield ratio must be stated, more than ever if the different bremsstrahlung endpoint energies are taken into account. This may be a consequence of the difficulties in determining this quantity in experiments with detectors, which cannot be operated during the beam pulse (s. section B.2.).

### C.2.2. (γ, 2n)-reactions

Only two isomers have been successfully investigated by (γ, 2n)-reactions so far. Fig. 11 shows the cross sections for $^{232}$U(γ, 2n) $^{236}$U and $^{239}$Pu(γ, 2n) $^{237m_1+m_2}$Pu (refs. [*16, 19, 20*]). The cross sections amount to ≈ 10 μb and 6 μb, respectively and are centered at about 17 MeV, corresponding to the higher (γ, 2n)-threshold. The reduction of the cross sec-

Table 3

Results for (γ, n)-reactions

| Reaction | $E_0$ (MeV) | $T_{1/2}$ | $Y_{iso}/Y_{pr.} \cdot 10^5$ | $\sigma_{iso}^{max}$(μb) | $E_{max}$ (MeV) | References |
|---|---|---|---|---|---|---|
| $^{235}U(\gamma, n)$ | 45 | < 10 ns | ≦ 0.01 | < 0.05 | 11.4*) | GÜNTHER et al. [21] |
| $^{240}Pu(\gamma, n)$ | 15.5 | 8.5 μs | | 170 ± 60 | 11.5*) | Gangrsky et al. [18, 43] |
| | 45 | (6.5 ± 0.4) μs | 7.9 ± 0.4 | 49 ± 3 | 12.3*) | GÜNTHER et al. [21] |
| $^{242}Pu(\gamma, n)\ ^{241m_1}Pu$ | 12.5 | (23 ± 1) μs | 115 ± 25 | 200 ± 60 | 10.5 | GANGRSKY |
| | 40 | (20.5 ± 2.5) μs | 9.2 ± 0.8 | 54 ± 5 | 12.0*) | et al. [18, 43] |
| $^{242}Pu(\gamma, n)\ ^{241m_2}Pu$ | 48 | (34 ± 7) ns | 3.7 ± 0.7) | | | GÜNTHER et al. [21] |
| $^{241}Am(\gamma, n)$ | 13 | 14 ms | 40 ± 10 | 150 ± 50 | 11.2 | GANGRSKY et al. [18] |
| | 41 | (14 ms) | 13 ± 1.5 | 60 ± 8 | 12.0 | KUZNETSOV et al. [23] |
| $^{243}Am(\gamma, n)$ | 12.5 | 0.9 ms | 70 ± 20 | 130 ± 50 | 10.6 | GANGRSKY et al. [18] |
| | 41 | (0.9 ms) | 18.5 ± 2.5 | 80 ± 10 | 12.0 | KUZNETSOV et al. [23] |

* The shape of the cross section was calculated (s. text).

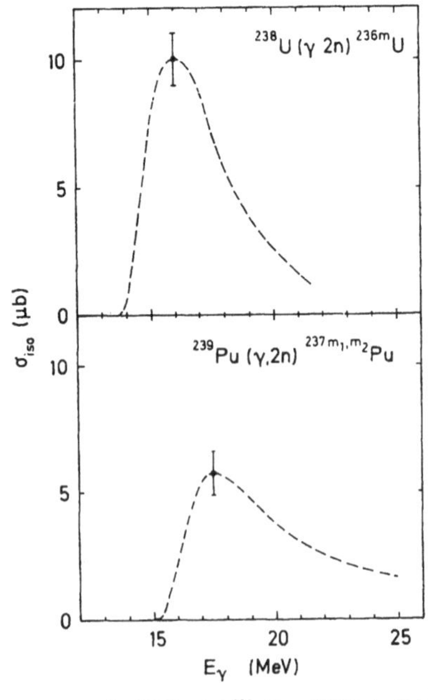


Fig. 11 Isomeric fission cross sections for $^{238}U(\gamma, 2n)\ ^{236m}U$ and $^{239}Pu(\gamma, 2n)\ ^{237m_1,m_2}Pu$ reactions [19, 21].

tions, compared with (γ, n)-reactions, can be explained in the $^{236}U$ case by a possible γ-decay of the isomer back to the first well. The (γ, 2n) cross section for $^{239}Pu$ is small because of the high fission probability $\Gamma_f/\Gamma_{tot}$ of this isotope. Therefore, a low population cross section of the $^{237}Pu$ shape isomers via a (γ, 2n) reaction is to be expected. The results for (γ, 2n) reactions obtained so far are summarized in table 4. The measured half-lives are in good agreement with data from particle induced reactions (s. ref. [*44*]). The systematics of the cross section maxima is discussed later on in section C.3.

Table 4
Results for (γ, 2n)-reactions

| Reaction | $T_{1/2}$ (ns) | $E_0$ (MeV) | $Y_{iso}/Y_{pr.} \cdot 10^5$ | $\sigma_{iso}^{max}$(μb) | $E_{max}$ (MeV) | References |
|---|---|---|---|---|---|---|
| $^{238}U$(γ, 2n) | (115 ± 5) | 45 | 2.02 ± 0.16 | 9.9 ± 1.0 | 16*) | GÜNTHER et al. [*19*] |
| | (118 ± 7) | | 2.10 ± 0.16 | 10.3 ± 1.0 | 16*) | GÜNTHER et al. [*16*] |
| $^{239}Pu$(γ, 2n) $^{237m_1}Pu$ | (77 ± 16) | 45 | 0.64 ± 0.17 | | 17.5*) | GÜNTHER et al. [*20*] |
| | (87 ± 11) | | 0.49 ± 0.07 | | 17.5*) | GÜNTHER et al. [*21*] |
| $^{239}Pu$(γ, 2n) $^{237m_2}Pu$ | (1050 ± 400) | 45 | 0.083 ± 0.022 | 5.7 ± 0.9 | 17.5*) | GÜNTHER et al. [*20*] |

*) The shape of the cross section was calculated (s. text).

### C.2.3. (γ, γ')-reactions

The only data available so far in literature come from GANGRSKY et al. [*18*] who reported on the population of the $^{239m_1}Pu$ fission isomer in a (γ, γ') reaction. Their results for the isomeric cross section, shown in fig. 12, amount to values of 8–10 μb for γ-energies between 8 and 10 MeV.

Unfortunately Gangrsky et al. could not measure directly the isomeric to prompt yield ratio $Y_{iso}/Y_{pr.}$. Using prompt fission cross sections [*36*] and a Schiff-bremsstrahlung

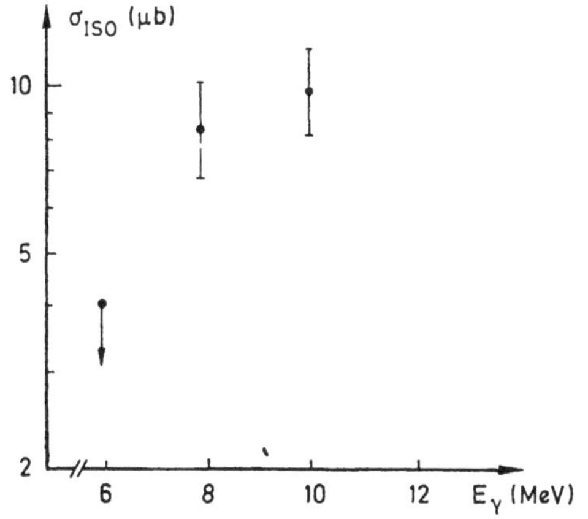


Fig. 12. Isomeric fission cross sections for the $^{239}Pu$(γ, γ') reaction reported by GANGRSKY et al. [*18*].

spectrum a yield ratio of

$$Y_{\text{iso}}/Y_{\text{pr.}}|_{E_{\text{BS}}=13\text{MeV}} \approx (1-2)\cdot 10^{-4}$$

can be derived from the cross section reported by Gangrsky et al. [*18*] depending slightly on the used extrapolation of the cross section up to 13 MeV (s. fig. 12). This result is challenged by recent attempts of the Giessen group to measure the population of $^{239m}$Pu in (γ, γ′)-reactions at a bremsstrahlungendenergy of 13 MeV. No isomeric fission could be detected. Their result of

$$Y_{\text{iso}}/Y_{\text{pr.}} < 4\cdot 10^{-6}$$

clearly shows that the cross sections must be at least one order of magnitude lower than those reported by Gangrsky et al. [*18*].

These lower cross sections seem to be reasonable when regarding, as crude guide, the cross sections for (γ, γ′) and (γ, n)-reactions in the first well [*36, 45*]. At an energy of ≈ 13 MeV these cross sections differ by a factor of ≈ 50—100. Since the isomeric fission cross sections for (γ, n)-reactions are about 50 μb (s. table 3) one should expect values < 1 μb for the population of fission isomers in (γ, γ′)-reactions.

Of special interest is a measurement of the (γ, γ′) isomeric fission cross section for $^{238}$U, since this would enable an estimation of the γ-decay of $^{238m}$U (back to the first minimum) from the Giessen results [*10*] of $^{238}$U(γ, xγ′) (s. section D.) First attempts at Giessen to measure this (γ, γ′) cross section even with the high sensitivity arrangement of a 30 stage parallel plate detector [*17*] (s. section B.2.3.) failed. Only an upper limit of

$$Y_{\text{iso}}/Y_{\text{pr.}}|_{12\text{MeV}} < 1.0\cdot 10^{-5}$$

could be detected corresponding to an isomeric fission cross section of $\sigma_{\text{iso}} < 1$ μb at about 10 MeV. This may be explained by both, the low population cross section of the isomer, and an overwhelming γ-decay of the isomer (s. section D.).

### C.3. Systematics of Isomeric Fission Cross Sections

As seen in the last chapter the isomeric fission cross sections in photonuclear reactions cannot be measured with good energy resolution as the excitation functions in particle induced reactions due to the use of the continuous bremsstrahlung. Therefore, in the following a restriction to a rather qualitative analysis is made to extract some information on the interesting barrier parameters from the measured cross sections.

In order to compare the isomeric photofission cross sections with those of particle induced fission and to get a systematic dependence on barrier parameters, a description proposed by Borggreen et al. [*46*] seems to be appropriate. A multistep process is assumed for the neutron evaporation and the subsequent fission.

For two neutron evaporation reactions (γ, 2n) on the nucleus $A+2$ the isomeric fission cross section can be written as:

$$\sigma_{\text{iso}}^{\text{max}} = \sigma_{\gamma,\text{abs}}^{A+2}(E_{\text{max}}) \times \left.\frac{\Gamma_{\text{n}}}{\Gamma_{\text{tot}}}\right|_{\text{I}}^{A+2} \times \left.\frac{\Gamma_{\text{f}}}{\Gamma_{\text{tot}}}\right|_{\text{I}}^{A+1} \times \left.\frac{\Gamma_{\text{n}}}{\Gamma_{\text{f}}}\right|_{\text{II}}^{A+1}. \tag{10}$$

The meaning of the width symbols can be easily seen in fig. 13, $\Gamma_{\text{n}}/\Gamma_{\text{f}}|_{\text{I,II}}^{A+1}$ describes the branching of a neutron or fission decay of the nucleus $(A+1)$ at the deformation of the first or second well.

For a (γ, n) reaction one can write in an analogous way:

$$\sigma_{\text{iso}}^{\text{max}} = \sigma_{\gamma,\text{abs}}^{A+1}(E_{\text{max}}) \times \frac{\Gamma_{\text{f}}}{\Gamma_{\text{tot}}}\bigg|_{\text{I}}^{A+1} \times \frac{\Gamma_{\text{n}}}{\Gamma_{\text{f}}}\bigg|_{\text{II}}^{A+1} . \tag{11}$$

$\Gamma_n/\Gamma_f|_{II}$ branchings can be derived from the measured $\sigma_{\text{iso}}^{\text{max}}$ data using known experimental $\Gamma_n/\Gamma_f|_I$ values. The parameters used and the results of such an analysis are summarized in table 5.

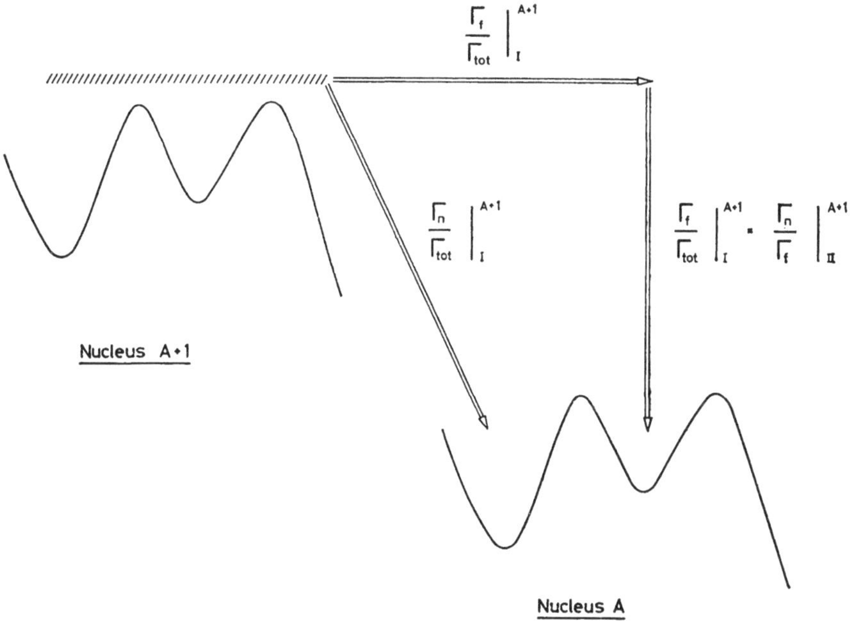


**Fig. 13.** Explanation of width symbols describing the population of class I and II state in the nucleus *A*

Table 5

$\Gamma_n/\Gamma_f|_{II}$ values and parameters used in the analysis

| Isomer A | $^{234}$U | $^{236}$U | $^{237m_1,m_2}$Pu | $^{239m_1}$Pu | $^{241m_1,m_2}$Pu | $^{240}$Am | $^{242}$Am |
|---|---|---|---|---|---|---|---|
| $\sigma_{\gamma,\text{abs}}^{A+X}(E^{\text{max}}_{-})$ (mb) | 268[b] | 120[b] | 160[d] | 470[d] | 470[d] | 470[d] | 470[d] |
| $\Gamma_n/\Gamma_f|_I^{A+2}$ | — | — | 0.25[b]<br>0.40[e] | — | — | — | — |
| $\Gamma_n/\Gamma_f|_I^{A+1}$ | 1.45[b] | 2.35[b] | 0.25[b]<br>0.3[e] | 0.54[e] | 0.88[b] | 0.56[g] | 0.62[g] |
| $E_B$ (MeV) | 6.15[c] | 5.7[e] | 5.4[e] | 5.42[f] | 5.5[h] | 5.2[e] | 4.95[e] |
| $E_{II}$ (MeV) | 2.6[c] | 2.3[m] | 2.3[m] | 2.4[f] | 2.0[e] | 2.7[e] | 2.4[i] |
| $S_n$ (MeV) | 6.18[k] | 5.90[k] | 6.58[k] | 6.18[k] | 5.87[k] | 6.41[k] | 6.10[k] |
| $\sigma_{\text{iso}}^{\text{pop}}$ | 59[a]*) | 64[a]*) | 7.2 ± 1.9[a]<br>5.7 ± 0.9[a] | 49 ± 3[a] | 51 ± 5[a] | 60 ± 8[g] | 80 ± 10[g] |
| $\Gamma_n/\Gamma_f|_{II}^{A+1} \cdot 10^4$ | < 0.003 | 1.0 ± 0.4 | 1.7 ± 0.6<br>1.35 ± 0.4 | 1.6 ± 0.1 | 2.2 ± 0.2 | 2.0 ± 0.3 | 2.8 ± 0.4 |

a) ref. [21]
b) ref. [36]
c) ref. [47]
d) ref. [40]
e) ref. [48]
f) ref. [49]
g) ref. [23]
h) ref. [50]
i) ref. [18]
k) ref. [52]
*) calculated s. D.

Following the concept of BORGGREEN et al. [46] the quantity $\Gamma_n/\Gamma_f|_{II}$ is plotted in fig. 14 as a function of $E_B - E_{II} - \tilde{S}_n$. $E_B$ is the height of the outer barrier, $E_{II}$ the energy of the second minimum and $\tilde{S}_n$ a mean neutron binding energy. The main uncertainty in $E_B - E_{II} - \tilde{S}_n$ arises from $E_{II} - E_B$. A total uncertainty of $\approx 0.3$ MeV is assumed except for $^{241}$Pu as discussed later on.
For comparison also $\Gamma_n/\Gamma_f|_{II}$ values from particle-induced fission are plotted in fig. 14 together with $\Gamma_n/\Gamma_f$ ratios for the groundstate deformation. Both ground-state and isomeric $\Gamma_n/\Gamma_f$ values are gathered around a straight line as given by a constant tempera-

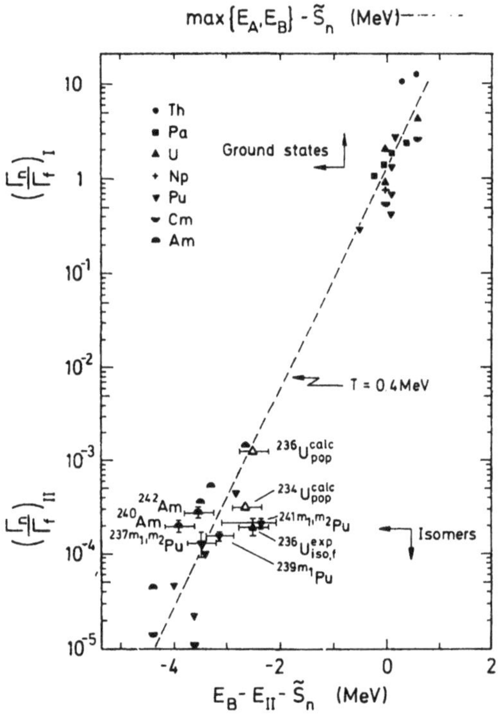


Fig. 14. Branching ratio $\Gamma_n/\Gamma_f/_{II}$ as a function of $E_B - E_{II} - \tilde{S}_n$ according to BORGGREEN et al. [46]. The photofission data are shown with error bars.

ture level density formula. The photofission results from Giessen [21] and the Am data of KUZNETSOV et al. [23] agree quite well with the isomer systematics except in the $^{236}$U and $^{241}$Pu case. This shows that generally the measured cross sections can be explained with the known barrier parameters.
The low value for $^{236}$U is quite reasonable in view of the $\gamma$-decay branch. Indeed the calculated population cross section (subsect. D.) fits very well the systematics. For $^{241}$Pu a $E_{II}$ value of 2.0 MeV is used as given by BRITT et al. [48]. However, regarding all experimental $E_{II}$ data for $^{241}$Pu there seem to be two groups of data centered around 2 and 2.7 MeV, respectively. A higher $E_{II}$ would improve the agreement of the photofission cross section with the $\Gamma_n/\Gamma_f|_{II}$ systematics.

## D. γ-Decay of Shape Isomers

### D.1. Introduction

For the lighter actinides ($Z < 92$) no fission isomers have been observed so far. This can be explained by the lowered inner barrier for these isotopes, leading to a favoured γ-decay of the shape isomers back to the first minimum of the potential energy surface. This supposition is supported by a great deal of experimental and theoretical work such as the systematics of the fission barriers [*47*–*51*] and of the half-lives of shape isomers [*3*], the interpretation of fission fragment angular distributions from photo [*53*] – and neutron-induced fission [*54*] and, in particular, the extremely low isomeric yields for $^{238m}$U (ref. [*55, 56*]), $^{236m}$U (ref. [*19, 27, 55*]) and $^{237m}$Np (ref. [*57, 58*]) as compared to the yields of the Pu-isotopes [*56*].
A reliable detection of a γ-decay of shape isomers provides important information on the penetrability through the lower inner barrier, which are not accessible to particle induced direct reactions. Moreover the γ-spectroscopic measurements represent a very accurate determination of the energy $E_{II}$ of the second well of the potential energy surface. The problems are mainly of experimental nature. First of all the detection efficiency of available γ-spectrometers is much lower than that of fragment detectors and, furthermore, there is a high background from delayed γ-quanta from fission fragments in such experiments. Photonuclear reactions seem to be favourable for this search for a γ-decay mode of shape isomers, since the population of spin isomers in residual nuclei and fission fragments is reduced due to the low spin transfer of these reactions.

### D.2. Statistical Model Estimations of the γ-Branch

An indirect estimate of the branching ratio $\Gamma_\gamma/\Gamma_f|_{II}$ for the decay of a shape isomer can be performed by comparing the experimentally observed cross section for isomeric fission with a cross section for the total production of the isomer, calculated e.g. in a statistical model. $\Gamma_{\gamma II}$ represents the width for a γ-decay of the shape isomer back to the first minimum and $\Gamma_{fII}$ that for isomeric fission.
This method has been applied to the photonuclear reaction $^{238}$U(γ, 2n) $^{236m}$U in ref. [*19*]. The energy dependence of the isomer population cross section $\sigma_{iso}^{pop}$ has been calculated by VANDENBOSCH's model [*35*] (see C.1.2.) for two neutron evaporation:

$$\sigma_{iso}^{pop} \sim \sigma(\gamma, 2n) \cdot \frac{\varrho_{II}}{\varrho_I} \cdot \frac{I(\Delta_2{}^1, 3) - I(\Delta_3, 3)}{I(\Delta_2, 1) - I(\Delta_3, 3)}. \tag{12}$$

For the calculation the absolute (γ, 2n) cross sections of the Livermore group [*36*] has been used. The absolute scale of $\sigma_{iso}^{pop}$ has been adjusted according to a consideration of ANDERSEN et al. [*59*] applied in the case of $^{235}$U(n, γ). In the most important energy region near the barrier of $^{236}$U the ratio $\sigma_{iso}^{pop}/\sigma_I$ of the populations into the two wells can be expressed in terms of the corresponding widths (see fig. 15)

$$\frac{\sigma_{iso}^{pop}}{\sigma_I} = \frac{\Gamma_2{}^\gamma \cdot \vec{\Gamma}_1}{\Gamma_1{}^\gamma \cdot \Gamma_2{}^{tot}} \quad (\Gamma_2{}^{tot} = \Gamma_2{}^\gamma + \Gamma_2{}^n + \vec{\Gamma}_2 + \overleftarrow{\Gamma}_2). \tag{13}$$

The ratio of the γ-widths is given by

$$\frac{\Gamma_2{}^\gamma}{\Gamma_1{}^\gamma} = \frac{(E - E_{II} - \Delta)^2}{(E - \Delta)^2} \tag{14}$$

where $E$ is the excitation energy and $\Delta$ the pairing energy of $\approx 1$ MeV.

Introducing the number of open channels $N_A$ and $N_B$ at the inner and outer barrier respectively we get

$$\frac{\sigma_{\text{iso}}^{\text{pop}}}{\sigma_{\text{I}}} = \frac{(E - E_{\text{II}}' - \Delta)^2}{(E - \Delta)^2} \cdot \frac{N_A}{N_A + N_B} \cdot \frac{\varrho_{\text{II}}}{\varrho_{\text{I}}}. \tag{15}$$

$\varrho_{\text{II}}/\varrho_{\text{I}} = 250$ can be taken from experiments of SOOD et al. [60]. Assuming $N_A = N_B$ for the nearly equal barriers in $^{236}U$ and for the low spin states populated in a ($\gamma$, 2n) reaction one can adjust the cross section $\sigma_{\text{iso}}^{\text{pop}}$ at an excitation energy of 6.4 MeV in $^{236}U$

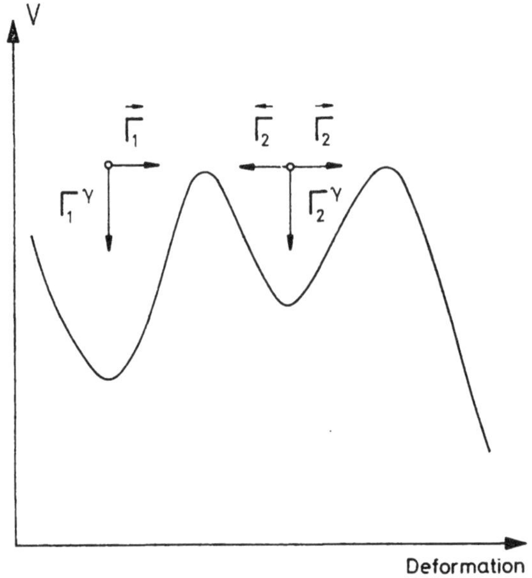


Fig. 15. Widths describing the coupling between class I and II states.

(neutron binding energy, corresponding to $E_\gamma \approx 17$ MeV). The so calculated cross section has a maximum of $\approx 65$ μb at 16 MeV and amounts to an integrated value of 208 μb × MeV.

The calculated integrated isomer population cross section of 208 μb · MeV must be compared with the measured isomeric fission cross section of (32 ± 6) μb · MeV. The discrepancy can be attributed to a competing γ-decay mode back to the first minimum. With this assumption one can deduce a branching ratio for the isomeric decay of

$$\Gamma_\gamma/\Gamma_f|_{\text{II}} \approx 6. \tag{16}$$

The value for the $^{236m}U$ branching ratio is in agreement with a similar estimation of ANDERSEN et al. [59] ($\Gamma_\gamma/\Gamma_f|_{\text{II}} = 7 \pm 2$), based on $^{235}U(n_{th}, f)$ experiments.

The same order of magnitude firstly was estimated by PEDERSEN et al. [61] by comparing the measured isomeric fission cross section in a $^{235}U$(d, p) reaction with a calculated population cross section.

The same statistical procedure has been applied to the $^{235}U(\gamma, f)\,^{234m}U$ reaction [21]. Comparing the calculated population cross section with the upper limit $Y_{\text{iso}}/Y_{\text{pr.}} < 10^{-7}$ for an isomer in $^{234}U$ with $T_{1/2} > 10$ ns as measured in Giessen [21] a branching ratio of $\Gamma_\gamma/\Gamma_f|_{\text{II}} > 1200$ can be deduced. However, it seems to be more probable that the total lifetime of the shape isomer in $^{234}U$ is drastically reduced by a predominant γ-decay of the isomer.

### D.3. Experimental Results for $^{238m}$U

Attempts of Borggreen et al. [46] to detect a γ-decay of $^{236m}$U and $^{238m}$Np via the $^{235}$U(d, p) and $^{238}$U(p, n) reactions failed because of insufficient sensitivity. The first report on the detection of a shape isomeric γ-decay came from Russo et al. [9], who carried out a (d, pn)-experiment on $^{238}$U. From the large number of observed delayed γ-lines most of them could be assigned to isomers in fission fragments by using different reactions and bombarding energies. Only two lines of 2.514 and 1.879 MeV remained, which have been attributed to the decay of the shape isomer to the $2^+$ rotational state (45 keV) and the $1^-$ state (680 keV) in the first minimum of $^{238}$U. A branching ratio for the decay of the shape isomer by γ-emission of $\Gamma_\gamma/\Gamma_f|_{II} \approx 20$ could be determined from the two observed transitions. An estimation of the strength of unobserved γ-lines increases this ratio by a factor of about 2. These values for the branching ratio can be explained by reasonable barrier parameters. The deduced barrier curvatures are somewhat higher than those derived from direct reaction studies. However, this may be due to the different range of excitation energy and consequence of the parabolic approximation of the barrier shape.

### D.4. Experimental Results for $^{236m}$U

There has been a preliminary report from the Heidelberg-group [62] on a possible γ-decay of the $^{236}$U shape isomer. In a $^{235}$U(d, p)experiment a weak line ($E = 2215$ keV; $T_{1/2} = (104 \pm 19)$ ns) has been observed, which has been tentatively interpreted as a γ-transition back to the first minimum. A branching ratio $\Gamma_\gamma/\Gamma_f|_{II}$ of $\approx 5$ has been evaluated in reasonable agreement with indirect, model dependent estimations [19, 59, 61].

This result has been challenged by a $^{238}$U(γ, 2n) $^{236m}$U experiment at the Giessen linac [10]. As mentioned earlier, the low transfer of angular momentum in this photonuclear reaction handicaps the population of spin isomers. Therefore the number of delayed γ-transitions is reduced.

The experiment has been performed at the bremsstrahlung facility of the Giessen linac (see fig. 3) using pulsed beam techniques. The γ-quanta from the Uranium target and from fission fragments were detected by a large volume Ge(Li)-diode (100 cm$^3$). The γ-events were analyzed with respect to their energy and delay time to the beam burst. The data were accumulated to energy-time-spectra, using the special two parameter measuring system with a $\log_2$ time scale (see B.3.).

Fig. 16 shows a section of an energy time matrix accumulated during three weeks. The γ-lines in the long-lived part of the spectrum come predominantly from γ-transitions in residual nuclei populated in β-decays of primary fission fragments. Most of the short-lived γ-activities (summarized in table 6) are caused by n-capture or n-scatttering on the Ge material of the γ-detector. Three γ-rays (642, 942 and 966 keV) belong to delayed transitions from the two quasi particle isomer in the first minimum of $^{236}$U (ref. [63]). The lines at 1181, 1280 and 1313 keV are transitions from high spin isomers in Te and Xe fission fragments [64]. The measured half-lives of all these transitions are in agreement with data from literature (see table 6). The 2225 keV γ-line originates from the n-capture by protons in the paraffin shielding around the γ-detector. The well known energy of this capture γ-ray (binding energy of the deuteron) was used for an additional "on-line"-calibration and as a stability control of the γ-spectrometer. In the interesting energy region between 2 and 3 MeV, where a shape isomeric γ-transition should be expected, no significant γ-line could be detected.

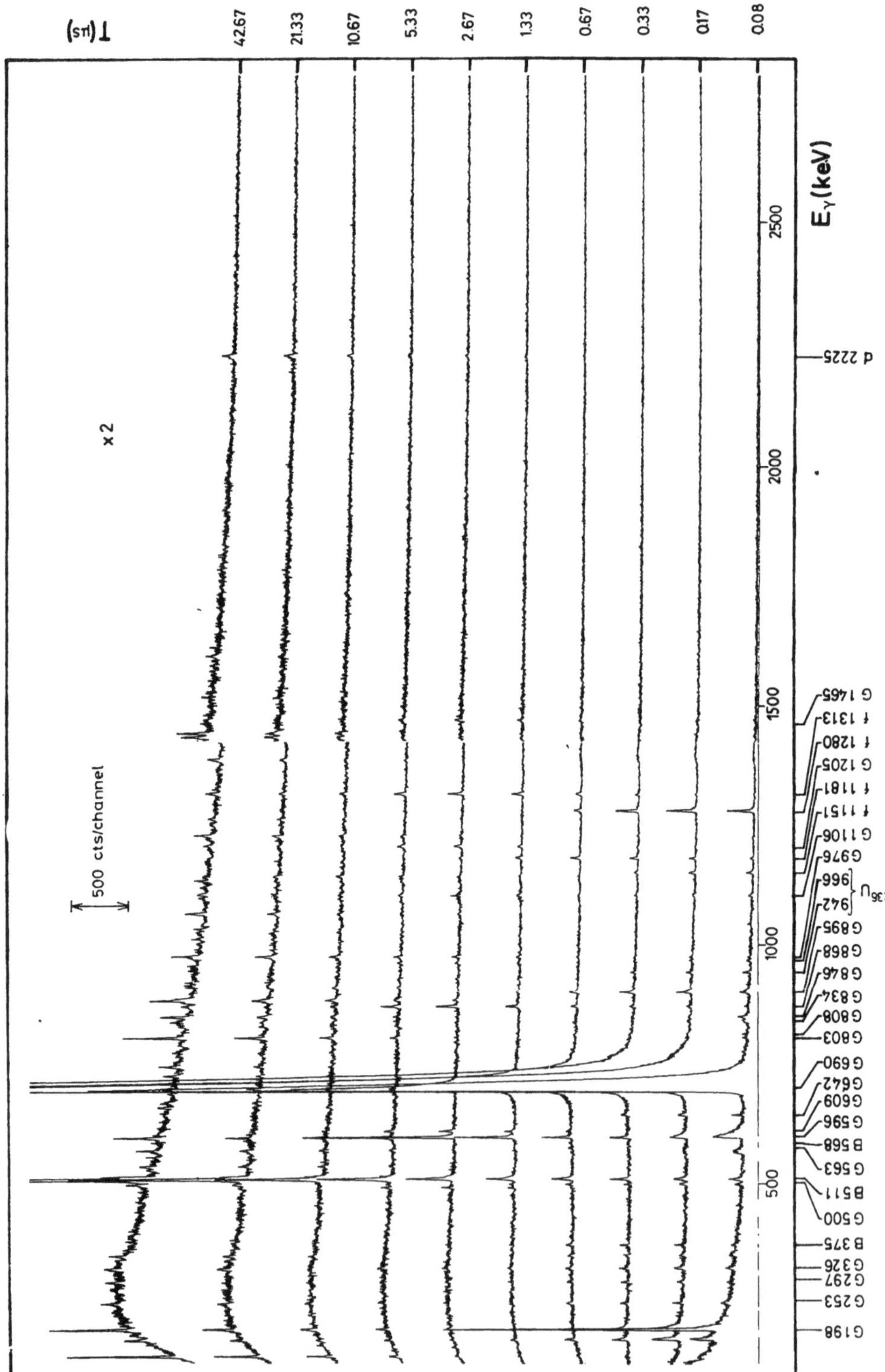


Fig 16. Energy-time-spectra measured in $^{238}$U($\gamma$, $x\gamma'$) (Ref. [10]).

Table 6
Short-lived γ-transitions observed in $^{238}$U(γ, $x$γ′)

| $E_\gamma$ (keV) | $t_{1/2}$ (ns) this work | $t_{1/2}$ (ns) literature | $I_{rel}$ (%) | Origin |
|---|---|---|---|---|
| 500 | $s$[b]), 2600[c]) | | | 70 Ge(n, γ) |
| 511 | | | | e$^+$-annihilation |
| 563 | $s$ | | 113 | 76 Ge(n, n′γ) |
| 568 | $s$, $l$[d]) | | | ? |
| 596 | $s$, 2600[c]), $l$ | | 423[e]) | 74 Ge(n, n′γ) |
| 609 | 3800 + 700 | | 262 | |
| 642 | 134 ± 35 | 130[i]) | 55 | $^{236m}$U(4$^-$) |
| 690 | 500 – 600[c]) | | ≈ 10$^5$ | $^{72}$Ge(n, n′IT) |
| 803 | $s$ | | 7 | ? |
| 808 | 99 ± 61 | | 13 | $^{73}$Ge(n, γ) |
| 834 | $s$ | | 13 | $^{72}$Ge(n, n′γ) |
| 846 | $s$ | | 17 | ? |
| 868 | 2600 ± 200,1 | | 182 | $^{74}$Ge(n, n′γ) |
| 895 | 281 ± 27 | | 113 | $^{72}$Ge(n, n′γ) |
| 942 | 120 ± 40 | 130[i]) | 17 | $^{236m}$U(4$^-$) |
| 966 | 73 ± 49 | 105[i]) | 8 | |
| 976 | 147 ± 117 | | 7 | $^{72}$Ge(n, γ) |
| 1106 | 101 ± 33 | | 11 | $^{73}$Ge(n, γ) |
| 1151 | 118 ± 20 | 79 ± 12[h]) 90[m]) | 15 | $f$[g]), $A$ ≈ 134 |
| 1181 | 641 ± 64 | 582 ± 99[h]) 670[m]) 590 ± 50[n]) | 70 | $^{135}$Te, $f$ |
| 1205 | 3900 ± 800 | | 44 | $^{73}$Ge(n, n′γ) |
| 1280 | 166 ± 6 | 154 ± 20[h]) 164[m]) 180 ± 50[n]) | 100 | $^{134}$Te, $f$ |
| 1313 | 3180 ± 240 | 3350 ± 469[h]) 3000[m]) 3000 ± 300[n]) | 95 | $^{136}$Xe, $f$ |
| 1465 | 4100 ± 2200 | | 14 | $^{72}$Ge (n, n′γ) |
| 2225 | 62000 ± 11000 | | 79 | p(n, γ) |

a) intensities normalized to $I_{1280}$ (expected value of $I^u_{\gamma,\mathrm{exp}} < 1\%$)
b) $s$ = short lived: $t_{1/2} < 100$ ns, intensities for an adopted $t_{1/2} = 100$ ns
c) estimated $t_{1/2}$ value
d) $l$ = long lived: $t_{1/2} > 1$ ms
e) $I$ for short lived component only
g) $f$ = fission fragment isomer
h) ref. [64]
i) ref. [63]
m) ref. [65]
n) ref. [66]

The low background in the energy-time spectra enabled an estimate to be made for the branching ratio $\Gamma_\gamma/\Gamma_f|_{II}$ of the decay of the $^{236}$U shape isomer. Two independent methods have been employed. The intensity for a shape isomeric γ-tansition was estimated first from the experimental intensity of the short-lived γ-transition in $^{134}$Te (1280 keV) and secondly from an experimental determination of the solid angle and the absolute photo-efficiency of the γ-spectrometer using the $^{24}$Mg(γ, pγ′)-reaction where absolute bremsstrahlung weighted cross sections are known [67]. Both evaluations resulted in equal intensities for the shape isomeric decay (see ref. [10]). An upper limit of $\Gamma_\gamma/\Gamma_f|_{II}$ could be deduced from the spectra due to the low background. Assuming that a γ-line with an intensity greater than two standard deviations of the background should be detectable, $\Gamma_\gamma/\Gamma_f|_{II}$ has been evaluated as a function of the γ-energy. Fig. 17 shows that the sensi-

tivity of this experiment gives an upper limit for $\Gamma_\gamma/\Gamma_f|_{II}$ of about 1.5 for a single high energetic γ-transition (2—3 MeV) from the $^{236}$U shape isomer.
These results are contradictionary to the preliminary Heidelberg report of $\Gamma_\gamma/\Gamma_f|_{II} \approx 5$ and also to the statistical model estimations [*19, 59*] and furthermore to the expectations from the systematics of the double humped fission barrier [*50, 51*]. A possible explanation of the low $\Gamma_\gamma/\Gamma_f$ ratio might be, that the total γ-strength is attributed to several low energy γ-transitions and is not concentrated in one or two high energetic transitions.

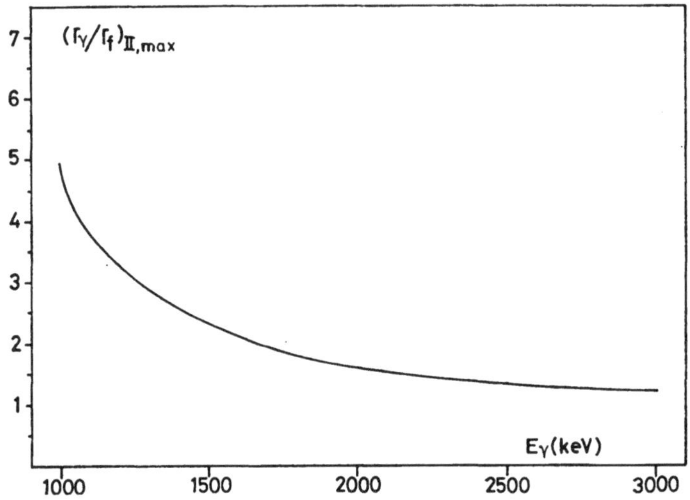


Fig. 17. Upper limit of the branching ratio $\Gamma_\gamma/\Gamma_f|_{II}$ for a single high energy isomeric γ-transition (Ref. [*10*]).

A confirmation of this low $\Gamma_\gamma/\Gamma_f$ branching ratio for a single γ-transition came very recently from a $^{235}$U(d, p) experiment performed by Butler et al. [*11*]. The most important improvement in this experiment was the use of a large NaJ/Tl crystal around the Ge(Li) detector leading to a reduction of the number of random coincidences. In this delayed coincidence (d, p)-experiment an additional signature for a shape isomeric γ-decay, besides γ-energy and isomeric half-life, should be an excitation energy dependence corresponding to the population cross sections of the isomer in a (d, p)-reaction [*68*]. Butler et al. [*11*] failed in detecting a delayed high energetic γ-line consistent with the $^{236m}$U half-life. From their spectra the authors derived an upper limit for $\Gamma_\gamma/\Gamma_f|_{II}$ using the $^{134}$Te intensity as in the photonuclear work [*10*]. The results shown in fig. 18 confirm the photonuclear data [*10*]. The deduced upper limit is plotted together with the photonuclear results and the value reported by Habs et al. (Heidelberg) [*62*].
The improved sensitivity of Butler's experiment for low γ-energies challenges the model dependent estimations of the total branching ratio $\Gamma_\gamma/\Gamma_f$. The nearly equal barriers of $^{236}$U and $^{238}$U as measured in direct reactions and the similar level scheme in the first minimum would suggest similar decay properties of the two isomers. The experimental results, however, of $\Gamma_\gamma/\Gamma_f|_{II} < 1.5(^{236}\mathrm{U})$ and $\Gamma_\gamma/\Gamma_f|_{II} \approx 20(^{238}\mathrm{U})$ (ref. [*9*]) differ quite considerably. Therefore an independent investigation of the $^{238m}$U decay should be performed to confirm this unexplained different behaviour of the decay of the $^{236m}$U and $^{238m}$U shape isomers. Unfortunately the $^{238}$U(γ, *x*γ′) experiment of the Giessen group [*10*] cannot be analyzed with respect to a γ-decay of the $^{238}$U isomer, since no $^{238}$U(γ, γ′) $^{238m}$U cross section is known so far (s. section C.2.3.).

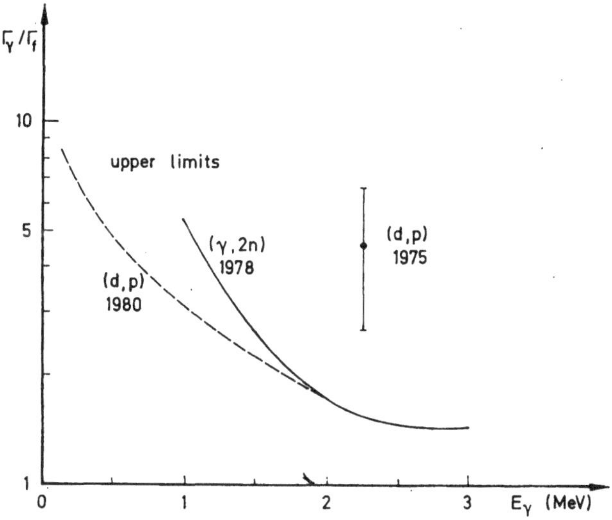


Fig. 18. Summarized results for $\Gamma_\gamma/\Gamma_f|_{II}$ of $^{236m}U$ (Refs. [*10*, *11*, *62*]).

## E. Spin Assignments from Isomeric Cross Section Ratios

### E.1. Excitations in the Second Well of the Potential Energy Surface

The investigation of excited states in the second minimum of the potential energy surface is of fundamental interest since more detailed information concerning the strongly deformed nucleus can be extracted from such experiments. Besides collective bands, low lying single particle excitations are possible. The identification and classification of single particle levels in the second well represent a crucial test of single particle models and their extrapolations to large deformations and consequently of the whole Strutinsky procedure.

Most of the fission isomers are ascribed to the ground state in the second well. Additional isomers are known in about ten isotopes [*3*]. In even-even nuclei these states are explained as two quasi-particle states [*69*] with an excitation energy of about 1.3 MeV analogous to the level scheme in the first well. In even-odd isotopes these second isomers are interpreted as low lying single particle excitations in the second minimum [*13*, *70*], the electromagnetic decay of which is hindered by spin selection rules. The spectroscopy within the second well should provide interesting information on the spins and parities of the levels involved.

### E.2. Experimental Methods and Previous Results from Particle Induced Reactions

Using the recoil shadow technique [*7*] and conversion electron spectroscopy Backe et al. [*71*] succeeded in determining for $^{239}Pu$ the spin of the ground-state in the second well and excitation energies. The proposed level scheme in the second minimum of $^{239}Pu$ represents the most detailed information on excitations in the second well.

Earlier attempts to deduce spins and K-quantum numbers in the case of $^{237m_1,m_2}Pu$ were carried out by Specht et al. [*7*] who investigated the fragment angular distributions from the two isomers. Kalish et al. [*73*] tried to measure the *g*-factor of the

1.1 μs spin isomer in $^{237}Pu$ by the spin precession method. Both experiments are not reproducible because of great experimental difficulties in preserving the alignment of the recoil nucleus [7, 74].

Another approach to deduce spins of fission isomers is the measurement of the relative population of two isomers in the same isotope in reactions with different angular momentum transfer into the compound system [70, 20]. The isomeric cross section ratio can be analyzed within the conventional statistical model procedure of HUIZENGA and VANDENBOSCH [75, 76]. Photonuclear reactions represent a significant extension of the

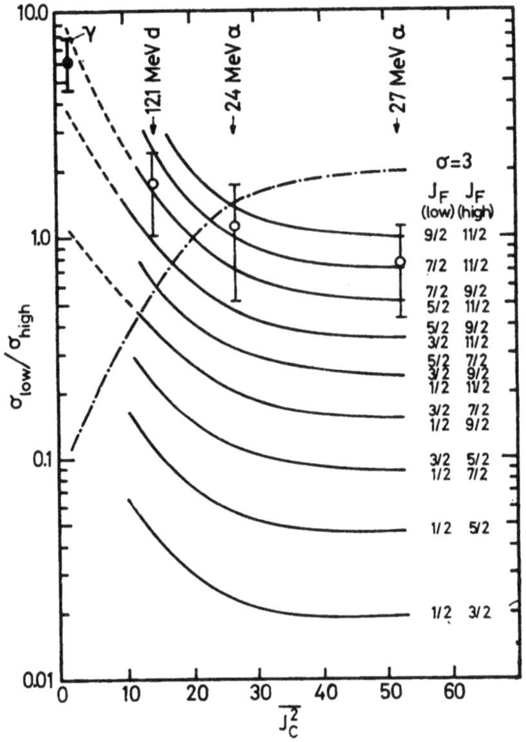


Fig. 19. Calculated isomeric yield ratios for different spin combinations in $^{237}Pu$ as a function of the mean squared spin of the initial compound nucleus.
Solid lines: Vandenbosch's calculations [70]
Dashed lines: Photonuclear extrapolations [20]
Dashed-dotted line: Isomeric ratio for inverse spin assignment (11/2–5/2)
Open circles: Results of (d, 2n) and (α, 2n) experiments [70]
Point: Result of (γ, 2n) experiment [20]

range of transferred angular momentum. The low spin transfer leads to a favoured population of low spin states and enables a more reliable identification of the high spin state. This is clearly demonstrated in fig. 19 (ref. [20]). The ratio of the population of the two isomers in $^{237}Pu$ is plotted, as calculated within the statistical model procedure, as a function of the mean squared angular momentum $\overline{I_c^2}$ of the compound nucleus for different spin combinations for the two states. The results for $^{237}Np(d, 2n)$ and $^{235}U(\alpha, 2n)$ reaction studies [70] are shown as open circles. The point represents the photonuclear ratio (γ, 2n)-reaction (ref. [20, 21]). The high sensitivity of photonuclear reactions for such an analysis is obvious regarding the dashed dotted line, representing the calculations for an inversed assignment of high and low spin to the two isomers. This method of deducing spin assignments from isomeric cross section ratios has been used for the investigation of the double isomers in the even-odd Pu-isotopes [*20, 22, 70*], in particular the comparison of particle induced reactions (high spin transfer) and photonuclear reactions (low spin transfer) should enable more reliable spin assignments and a test of different single particle calculations.

### E.3. Statistical Model Analysis of Photonuclear Isomeric Ratios

HUIZENGA and VANDENBOSCH [*75, 76*] have developed a statistical model procedure to calculate isomeric cross section ratios and their energy dependence. Fitting the calculated isomeric ratios to the experimental ones it is possible to obtain information on the spin dependence of the nuclear level density, in particular on the spin cut-off parameter $\sigma$ (SCOP), and the level density parameter $a$. In the past these quantities were often deduced from the measured cross section ratios. On the other hand, the Huizenga-Vandenbosch method can be used to calculate excitation functions of isomeric states in order to get information about the spin assignment of the isomeric state.
The Huizenga-Vandenbosch model (HVM) was often used, sometimes in a modified way, to calculate isomeric cross section ratios for nuclei which were produced by different reactions. Therefore only a short summary of the most important steps and assumptions of the HVM for the special case of photonuclear reactions is presented.

(i) Starting with a target nucleus with spin $J_0$, the absorption of an $E_1$ $\gamma$-ray leads to excited states in the first compound system $A^*$ with spins $J_c = J_0, J_0 \pm 1$. For $J_0 = 0$ only states with $J_c = 1$ are excited. The relative occupation probability is proportional to the number of magnetic substates

$$P(J_c) \sim 2J_c + 1\,. \tag{17}$$

(ii) The spin distribution in the first compound system $A^*$ is changed by the successive evaporation of one or more neutrons. The neutron transition probability from a state with spin $J_i$ to a state with spin $J_f$ is given by

$$P(J_f) \sim \varrho(J_f)\, P(J_i) \sum_{S=|J_f-1/2|}^{|J_f+1/2|} \; \sum_{l=|S-J_i|}^{|S+J_i|} T_l(E_n) \tag{18}$$

where $T_l(E_n)$ is the penetrability for a neutron with angular momentum $l$ and kinetic energy $E_n$. The evaporation energy $E_n$ is replaced by an averaged value $\bar{E}_n$, assumed to be

$$\bar{E}_n = t \tag{19}$$

with $t$ calculated from

$$U = at^2 - t \tag{20}$$

where $U$ is the excitation energy and $t$ the thermodynamic temperature.

(iii) If the remaining excitation energy is below the particle threshold it is supposed that the residual nucleus deexcitates predominantly by $E1$ $\gamma$-ray emission with an average energy of

$$E_\gamma = 4\big((U/a) - (5/a^2)\big)^{1/2} \tag{21}$$

where $U$ is the residual excitation energy. The $\gamma$-ray transition probability to states with spin $J_f$ is assumed to be

$$P(J_f) \propto \sum_{J_i} P(J_i)\, \varrho(J_f)\,. \tag{22}$$

(iv) The γ-cascade is continued until the residual energy is smaller than the "γ-ray cut-off" energy. The following γ-ray, the "deciding gamma", for which an arbitrary multipolarity is admitted now, feeds that state to which the transition has the lowest multipolarity. If the residual energy remains within the "γ-ray cut-off" region, partly a further $E_1$ γ-ray is emitted and partly the "deciding gamma" is emitted as described above.
(v) The relative probability for the population of the isomeric pair is calculated by the following procedure: One calculates the "center of spins" (COS) for the two levels, defined by $COS = 1/2(J_{high} + J_{low})$. All states with spin $<$ COS feed the low-spin state, all states with spin $>$ COS feed the high-spin states and with spin $=$ COS feed both states in equal parts. If a further state exists between the ground state and isomeric state with an intermediate spin value, this level is considered to compete with the isomeric state instead of the ground state. The decision for the population of both states is made in a manner analogous to that described above. The average evaporation energy for the neutrons is assumed to be $E_n = t$ and the γ-ray cut-off boundaries to be 1 and 2 MeV. The neutron penetrabilities can be taken from the tables of AUERBACH and PEREY [77] and threshold energies from MATTAUCH et al. [52].

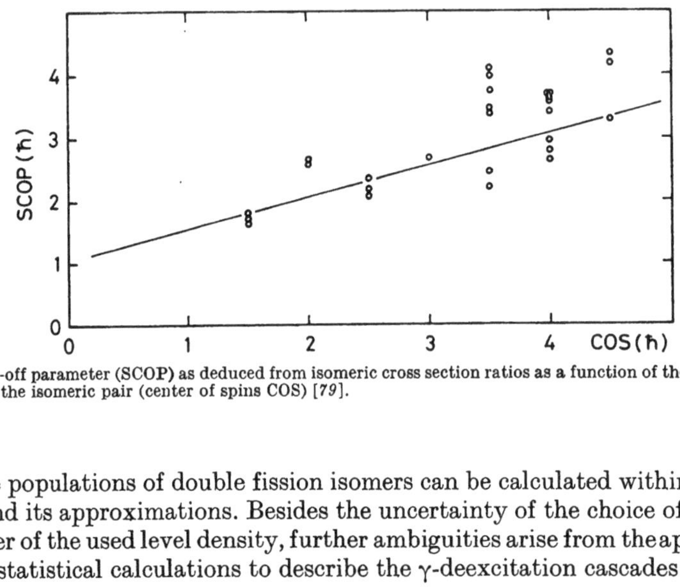


Fig. 20. Spin-cut-off parameter (SCOP) as deduced from isomeric cross section ratios as a function of the mean value of the spins of the isomeric pair (center of spins COS) [79].

The relative populations of double fission isomers can be calculated within this statistical model and its approximations. Besides the uncertainty of the choice of the spin cut-off parameter of the used level density, further ambiguities arise from the approximations used in the statistical calculations to describe the γ-deexcitation cascades after neutron evaporation.
Only $E1$ γ transitions were assumed, feeding states differing in spin by $0\hbar$ or $\pm 1\hbar$ from that of the original level. This crude approximation, neglecting collective $E2$ transitions toward lower spin states, clearly overestimates the population of high spin states.
Since in general no level schemes in the second well are known so far, no reasonable refinement of the deexcitation model seemed to be possible, i.e., $E2$ contributions could not be taken into account without introducing a new arbitrary parameter into the model. Therefore, a more phenomenological approach can be attempted. In the past a systematic analysis of all available photonuclear (γ, $xn$) isomeric cross section ratio measurements for isomers with known spins has been performed [78, 79]. This analysis showed a linear correlation between spin cutoff parameters ($\sigma$) derived by the simple statistical model [75, 76] and the mean value of the spins of the isomeric pair (COS) (s. fig. 20). If one uses, vice versa, this empirical correlation between $\sigma$ and the mean spin value for spin assignments, indirectly the shortcomings of the model are taken into account.

This procedure was applied to analyze the photonuclear isomeric yield ratios for the double isomers in $^{237}$Pu and $^{241}$Pu (refs. [*20, 22*]). Unfortunately the isomeric ratio procedure could not be tested on $^{239}$Pu with its known level scheme, since the short lived $^{239}$Pu spin isomer could not be observed in an unambiguous way in a photonuclear reaction [*21*].

### E.4. Results for $^{237m_1m_2}$Pu

One of the most investigated nucleus with two fissioning isomers in the second well is $^{237}$Pu ($T_{1/2} \approx 100$ ns and 1.1 μs respectively) (s. refs. [*20, 13*]). These Pu isomers can be populated with sufficient high cross sections in (α, 2n), (d, 2n) and (γ, 2n) reactions on available targets ($^{235}$U, $^{237}$Np, $^{239}$Pu). From excitation function measurements [*80*] it was concluded that the long-lived state lies about 0.3 MeV above the 100 ns level. Consequently the short-lived state was ascribed to the ground state in the second well and the 1.1 μs level is assumed to correspond to a single particle excitation of the odd neutron. Furthermore it is known in the case of $^{237}$Pu that the higher lying 1.1 μs isomer predominantly decays by fission rather than by an electromagnetic decay to states in the second well. This was very recently shown by the Heidelberg group [*81*] using a magnetic separator for isomeric recoil ions. Similar results were reported by H. H. Duhm et al. [*82*] from an $e_k - f$ delayed coincidence experiment. Also the older fragment angular distribution measurements of Specht et al. [*72*] seem to suggest the predominant fission decay of the higher lying spin isomer. Therefore no electromagnetic depopulation of the 1.1 μs isomer is taken into account in the following analysis.
In fig. 19 the results of (α, 2n), (d, 2n) and (γ, 2n) reaction studies were shown together with the statistical model calculations using a spin cut-off parameter $\sigma = 3\hbar$ for different spin combinations. The trend of the experimental data seems to be described by spin assignments 5/2—11/2, 7/2—9/2, 7/2—11/2. Unfortunately the yield ratio calculated within the statistical model formalism strongly depends on the adopted spin cut-off parameter. This can be seen in fig. 21 where the photonuclear yield ratio $Y_{low}/Y_{high}$ is plotted as a function of $\sigma$. Values of $3-5\hbar$ for $\sigma$ are regarded to be reasonable [*70*]. Four probable spin combinations can be derived from the comparison of the calculated yield ratio and the photonuclear result [*20, 21, 22*] for a spin cut off parameter range of $3-5\hbar$

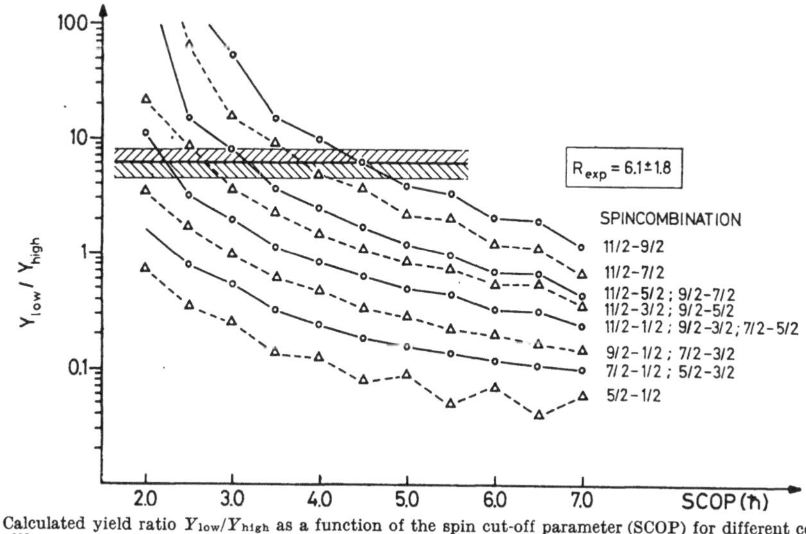

Fig. 21. Calculated yield ratio $Y_{low}/Y_{high}$ as a function of the spin cut-off parameter (SCOP) for different combinations ($^{237m_1,m_2}$Pu).

(11/2—9/2, 11/2—7/2, 11/2—5/2, 9/2—7/2). From lifetime arguments the spin pairs 9/2—7/2, 11/2—7/2, and 11/2—9/2 can be excluded as possible candidates from the isomeric pair in $^{237}$Pu. Owing to the small multipolarities of the $\gamma$ transitions for these spin combinations (dipole and quadrupole radiation), the $\gamma$-transition lifetime of the upper, high spin level would be too small as compared with the observed half-life of 1100 ns. Therefore, this first analysis provides as the most probable spin assignment the spin $I = 11/2$ to the long-lived, energetically higher isomer and $I = 5/2$ to the ground state in the second minimum of $^{237}$Pu.

The results of this analysis are only valid in the context of the statistical model and its approximations. Some of the drawbacks can be taken into account using empirical linear correlation between the mean spin value of the isomeric pair (COS) and the spin cut-off parameter $\sigma$ as discussed in the preceeding section.

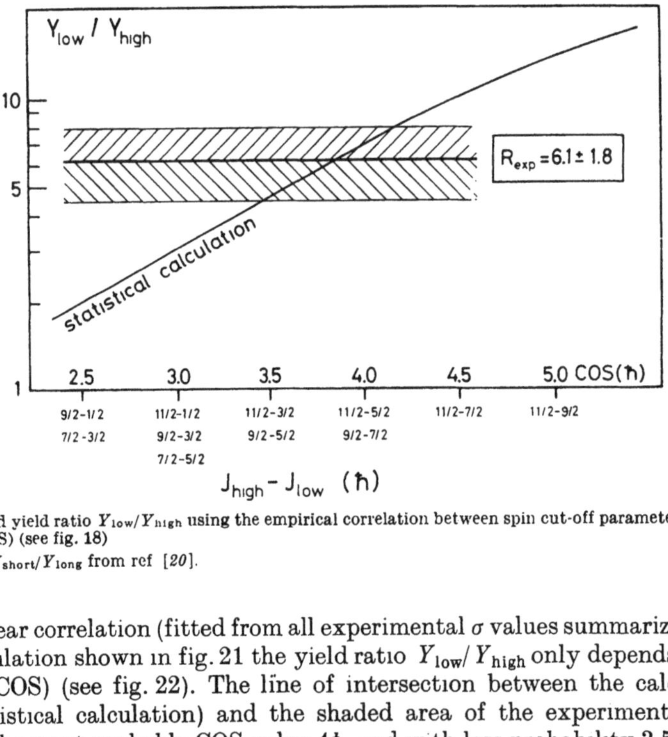


Fig 22. Calculated yield ratio $Y_{low}/Y_{high}$ using the empirical correlation between spin cut-off parameter and the center of spins (COS) (see fig. 18)
$R_{exp} = Y_{short}/Y_{long}$ from ref [20].

Using the linear correlation (fitted from all experimental $\sigma$ values summarized in ref. [79]) and the calculation shown in fig. 21 the yield ratio $Y_{low}/Y_{high}$ only depends on the mean spin value (COS) (see fig. 22). The line of intersection between the calculated curve (labeled statistical calculation) and the shaded area of the experimental yield ratio provides as the most probable COS value $4\hbar$, and with less probability $3.5\hbar$, which cannot be excluded definitely (see fig. 22). Since spin combinations with spin differences $\leq 2\hbar$ again can be excluded by lifetime arguments 11/2—5/2 remains as the most probable spin combination also in this modified analysis. This assignment fully agrees with the conclusion of VANDENBOSCH et al. [70] based on the ($\alpha$, 2n) and (d, 2n) isomeric ratios.

### E.5. Results for $^{241m_1,m_2}$Pu

The two fission isomers in $^{241}$Pu have been investigated recently in the $^{242}$Pu($\gamma$, n) reaction [21, 22] (s. table 3). From the measured relative population an isomeric yield ratio

$$R_{exp} = Y_{long}/Y_{short} = 2.5 \pm 0.6 \qquad (23)$$

could be derived.

The analysis in the case of $^{241}Pu$ is even more complicated because of the unknown branching ratio for the decay of the higher lying isomer. It may be concluded from the systematics of the half-lives of the fission isomers [3] that the long-lived isomer in $^{241}Pu$ corresponds to the ground-state in the second well and the short-lived isomer to a higher lying spin isomer. This isomer can decay either by electromagnetic transitions (conversion electrons or γ-decay) with a width $\Gamma_\gamma$, representing the sum of the decay widths to different lower lying states, or by fission with a width $\Gamma_f$. Defining the branching ratio $B$

$$B = \Gamma_\gamma^{\text{short}}/(\Gamma_\gamma^{\text{short}} + \Gamma_f^{\text{short}}) \tag{24}$$

one can write the experimental isomeric yield ratio $R_{\text{exp}}$ in the following way

$$R_{\text{exp}} = Y_{\text{iso}}^{\text{long}}/Y_{\text{iso}}^{\text{short}} = \frac{\sigma_{\text{iso,pop}}^{\text{long}} + B \cdot \sigma_{\text{iso,pop}}^{\text{short}}}{\sigma_{\text{iso,pop}}^{\text{short}} \cdot (1 - B)}. \tag{25}$$

$\sigma_{\text{iso,pop}}^{\text{long,short}}$ are the population yields of the long and short-lived isomers respectively. From equ. (25) follows:

$$\sigma_{\text{iso,pop}}^{\text{long}}/\sigma_{\text{iso,pop}}^{\text{short}} = R_{\text{exp}} - B(R_{\text{exp}} + 1). \tag{26}$$

The upper limit of the branching ratio $B = (0.71 \pm 0.05)$ is given by the nonpopulation of the long-lived isomer ($\sigma_{\text{iso,pop}}^{\text{long}} = 0$), using the experimental value $R_{\text{exp}} = 2.5 \pm 0.6$. There are two possibilities to be discussed. One can ascribe the low spin to the long-lived and the high spin to the short-lived isomer and vice versa. For both cases and each spin combination we calculated the branching ratio $B$ needed to reproduce the measured yield ratio $R_{\text{exp}}$ (see fig. 23). The solid line represents the branching ratio for an assignment of the higher spin to the ground-state and the lower spin to the short-lived spin isomer. In this case the ground-state transition of the short-lived spin isomer should be

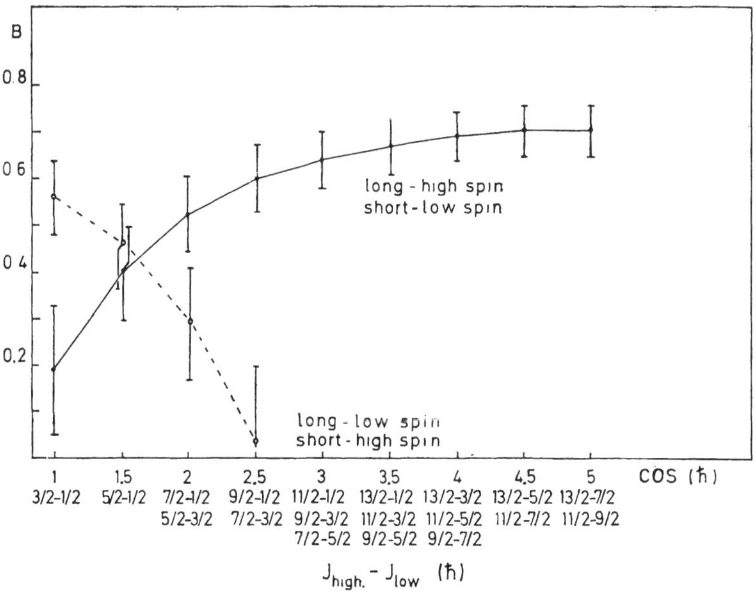


Fig. 23. Branching ratio $B$ as a function of the mean spin value (COS) for an assignment of the high spin to the long-lived isomer (full line) and the short-lived isomer (dashed line).

the dominating electromagnetic decay. In particular no population of the rotational band built on the ground state is to be expected due to the higher spin differences and lower transition energies. Therefore transitions with $\Delta J \leqq 1$ and $\Delta J \geqq 3$ and be excluded from lifetime arguments. The remaining combinations are summarized in table 7.

Table 7

Spin combinations and branching ratios for the assignment of the high spin to the ground state

| | 20 μs isomer | 34 ns spin isomer | $B$ (%) | comments |
|---|---|---|---|---|
| spin combination | 5/2 | 1/2 | $40 \pm 10$ | possible |
| | 7/2 | 3/2 | $60 \pm 7$ | unlikely |
| | 9/2 | 5/2 | $67 \pm 6$ | unlikely |
| | 11/2 | 7/2 | $70 \pm 5$ | unlikely |

Regarding the isomeric fission cross sections obtained in (d, p)-reactions for the long and short-lived isomer ((0.9 ± 0.4) μb [*56*] and (0.50 ± 0.09) μb [*83*], respectively) one gets an isomeric cross section ratio $R = 1.80 \pm 0.85$. This value deviates not significantly from the photonuclear ratio $R_{\text{exp}} = 2.5 \pm 0.6$. This is a strong hint that the spins of the two isomers should not differ very much. As can be seen in tab. 7 the last three combinations need a high branching ratio $B$ (60—70%) corresponding to no direct population of the ground-state in a photonuclear reaction. For a (d, p)-reaction with its larger transfer of angular momentum a higher population of the high spin state is expected, leading to an essentially higher ratio $R$. This was not experimentally observed, therefore the last three spin combinations are regarded as not very probable.

The dashed line in fig. 23 shows the branching ratio $B$ for the assignment of the high spin to the short-lived spin isomer and the low spin to the ground-state. In this case a decay of the spin isomer to the rotational band is to be expected, as observed by Backe et al. [*71*] for $^{239}$Pu. From lifetime arguments we can exclude only faster dipole transitions ($\Delta J = 1$). The following combinations remain (see table 8): (1/2—5/2), (1/2—7/2), (3/2—7/2), (1/2—9/2). The combination 1/2—9/2 with a branching ratio of only 3%

Table 8

Spin combinations and branching ratios for the assignment of the low spin to the ground state

| | 20 μs isomer | 34 ns spin isomer | $B$ (%) | comments |
|---|---|---|---|---|
| spin combination | 1/2 | 5/2 | $46 \pm 9$ | possible |
| | 1/2 | 7/2 | $29 \pm 12$ | possible |
| | 3/2 | 7/2 | $3 \pm 17$ | possible |
| | 1/2 | 9/2 | $3 \pm 17$ | unlikely |

fission decay seems to be very unlikely in view of the large spin difference and regarding the nearly equal (d, p)- and (γ, n)-yield ratios. However, the three other assignments 1/2—5/2, 1/2—7/2, 3/2—7/2 together with the combination 5/2—1/2 from tab. 7 have to be regarded as equivalent and no further restrictions can be made because of the unknown decay scheme in the second well.

## E.6. Spectroscopic Data for the Even-Odd Pu Isotopes and Comparison with Single Particle Calculations

In fig. 24 the available spectroscopic data for the even-odd Pu isotopes are summarized. The spin assignments for the fission isomers now can be compared with single particle shell model calculations to identify the levels as certain Nilsson orbits. Fig. 25 shows the calculations of several authors [*84*–*88*] using various single particle potentials.

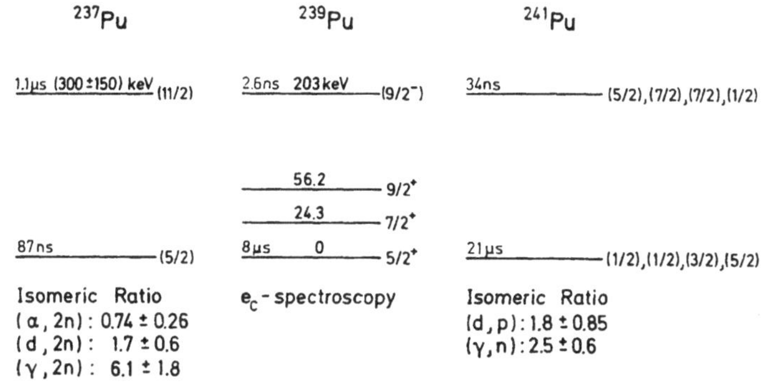


Fig. 24. Comparison of the spectroscopic information on the fission isomers in even-odd Pu-isotopes.

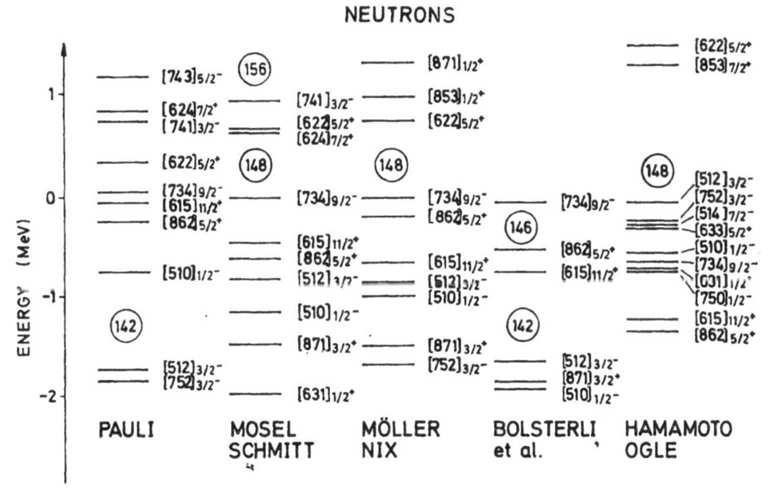


Fig. 25. Single particle energy levels for neutrons at a deformation corresponding to the second minimum of the potential energy surface. The figure is taken from R. VANDENBOSCH's review article [*13*] except the level scheme calculated by HAMAMOTO and OGLE [*84*]. This level scheme as constructed by V. METAG et al. [7] corresponds to slightly different deformation parameters describing the Heidelberg results [*71*] and changing the [633] and [862] levels (see comments to figs. 7 and 4 or refs. [*13*] and [7] respectively).

The results of all authors are quite similar besides the calculation of HAMAMOTO and OGLE [*84*], who used a stronger and deformation dependent spin-orbit force.
Very recently LIBERT et al. [*89*] calculated spectroscopic properties of the $^{239}$Pu fission isomers within the rotor plus quasi-particle model from self-consistently determined single particle states. The main components of the quasi-particle states in a harmonic oscillator basis are given. The single particle spectrum is similar to that calculated by

MÖLLER and NIX [87]. LIBERT et al. [89] succeeded in describing very well the energy spectrum and the quadrupole moment for $^{239m}$Pu and confirmed the spin and parity assignments of BACKE et al. [71].
For $^{237}$Pu ($N = 143$) both statistical model analysis [20, 70] favoured a spin assignments 5/2—11/2 for the ground-state and the spin isomer, respectively. Nilsson orbits $[615]^{11/2+}$ and $[862]^{5/2+}$ can be found near the Fermi surface in all calucations. Unfortunately, there are also $[622]^{5/2+}$ and $[633]^{5/2+}$ levels in the neighbourhood.
For $^{239}$Pu ($N = 145$) the situation is mostly clarified. BACKE et al. [71] determined a spin $5/2^+$ of the long-lived 8 μs isomer. From their results they concluded a contribution of more than 60% to the wave function of the 8 μs isomer having an antiparallel coupling of spin and angular momentum. Only in the Hamamoto calculation [84] a corresponding level $[633]^{5/2+}$ is present. This state, however, is strongly mixed with the $[862]^{5/2+}$ orbit.
As mentioned above all spectroscopic properties measured by BACKE et al. [71] could be reproduced by the calculations of LIBERT et al. [68] except the $E1/M1$ mixing ratio of the rotational transitions, leading to the [633] Nilsson classification of the 5/2 ground state.
Due to the unknown decay branching of the spin isomer the situation in $^{241}$Pu is very complicated as discussed above. For the most probable spin combinations namely: 1/2—5/2, 1/2—7/2, 3/2—7/2 and 5/2—1/2 corresponding Nilsson levels can be found in most of the calculations near $N = 147$. Assuming a $5/2^+$ ground-state as in $^{237}$Pu and $^{239}$Pu the spin of the 34 ns isomer should be 1/2, most probable a $[510]^{1/2-}$ orbit. In the calculations of MÖLLER et al. [87] and BOLSTERLI et al. [88] no 7/2 state appears near the Fermi surface. In the calculation of HAMAMOTO and OGLE [84] two 3/2 and two 7/2 states are presented exactly at the Fermi surface.
Further experiments, also for $^{243}$Pu, are needed to clarify the spin assignments in the even-odd Pu-isotopes in order to perform a more detailed systematic comparison or even a valuation of the different model calculations and their assumptions, in particular the deformation dependence of the spin orbit force.

## Acknowledgments

We are indebted to Prof. Dr. U. Mosel and Dr. V. Metag for stimulating discussions and helpful suggestions. Thanks are due to Dr. H. J. Maier for the excellent target preparations. We are grateful to Dr. H. Bartsch, Dipl. Phys. J. Drexler, Dipl. Phys. H. Ströher, H. Ries and W. Wilke for their active participation at different states of the experiments.
The financial support by the Deutsche Forschungsgemeinschaft is gratefully acknowledged.

## References

[1] S. M. POLIKANOV, V. A. DRUIN, V. A. KARNAUKHOV, V. L. MIKHEEV, A. A. PLEVE, N. K. SKOBELEV, V. G. SUBBOTIN, G. M. TER-AKOPJAN, V. A. FOMICHEV, Soviet Physics-JETP **15**, 1016 (1962).
[2] V. M. STRUTINSKY, Nuclear Phys. **A 95**, 420 (1967); V. M. STRUTINSKY, Nuclear Phys. **A 122**, 1 (1968).
[3] V. METAG, Nukleonika **20**, 789 (1975).
[4] H. J. SPECHT, J. WEBER, E. KONECNY, D. HEUNEMANN, Phys. Letters **41 B**, 43 (1972).
[5] D. HABS, V. METAG, H. J. SPECHT, G. ULFERT, Phys. Rev. Letters **38**, 387 (1977); G. ULFERT, V. METAG, D. HABS, H. J. SPECHT, Phys. Rev. Letters **42**, 1596 (1979).
[6] G. ULFERT, D. HABS, B. METAG, H. J. SPECHT, Nuclear Instr. Meth. **148**, 369 (1978).
[7] V. METAG, D. HABS, H. J. SPECHT, Phys. Reports **65**, 1 (1980).
[8] C. E. BEMIS Jr., J. R. BEENE, J. P. YOUNG, S. D. KRAMER, Phys. Rev. Letters **43**, 1854 (1979).

[9] P. A. Russo, J. Pedersen, R. Vandenbosch, Nuclear Phys. **A 240**, 13 (1975).
[10] H. Bartsch, W. Günther, K. Huber, U. Kneissl, H. Krieger, H. J. Maier, Nuclear Phys. **A 306**, 29 (1978).
[11] P. A. Butler, R. Daniel, A. D. Irving, T. P. Morrison, P. J. Nolan, V. Metag, J. Phys. G: Nuclear Phys. **6**, 1165 (1980).
[12] R. Vandenbosch, J. R. Huizenga, Nuclear Fission, Academic Press, New York (1973).
[13] R. Vandenbosch, Ann. Rev. Nuclear Sci. **27**, 1 (1977).
[14] V. Metag, Proceed. Internat. Symp. Phys. Chem. Fission, Jülich 1979, Vol. I, p. 153 IAEA Publ., IAEA-SM-241/B 3.
[15] D. N. Poenaru, Ann. Phys. (France) t **2**, 133 (1977).
[16] W. Günther, K. Huber, U. Kneissl, H. Krieger, S. Schulz, Nuclear Instr. and Meth. **163**, 459 (1979); W. Günther, Thesis, unpublished, Giessen (1980).
[17] H. Ries, J. Drexler, R. Fischer, W. Günther, K. Huber, U. Kneissl, H. Ströher, W. Wilke, Nuclear Instr. Meth. **185**, 373 (1981); H. Ries; Diploma Thesis, Giessen 1981, unpublished; W. Wilke; Diploma Thesis, Giessen 1980, unpublished.
[18] Yu. P. Gangrsky, B. N. Markov, Yu. M. Tsypenyuk, Fortschr. Phys. **22**, 199 (1974).
[19] W. Günther, K. Huber, U. Kneissl, H. Krieger, Nuclear Phys. **A 297**, 254 (1978).
[20] W. Günther, K. Huber, U. Kneissl, H. Krieger, H. J. Maier, Phys. Rev. **C 19**, 433 (1979).
[21] W. Günther, K. Huber, U. Kneissl, H. Krieger, H. J. Maier, H. Ries, H. Ströher, W. Wilke, Nuclear Phys. **A 350**, 1 (1980).
[22] W. Günther, K. Huber, U. Kneissl, H. Krieger, H. Ries, H. Ströher, W. Wilke, Nuclear Phys. **A 359**, 397 (1981).
[23] V. L. Kuznetsov, L. E. Lazareva, V. G. Nedorezov, N. V. Nikitina, N. M. Parovic, Nuclear Phys. **A 324**, 29 (1979).
[24] V. Metag, E. Liukkonen, G. Sletten, O. Glomset, S. Bjørnholm, Nuclear Instr. Meth. **114**, 445 (1974).
[25] H. Beil, R. Bergère, CEA-N-2144 (1980), CEN-Saclay (France).
[26] B. Tamain, B. Peiffer, H. Wollnik, E. Konecny, Nuclear Phys. **A 173**, 465 (1971).
[27] J. Christiansen, G. Hempel, H. Ingwersen, W. Klinger, H. Schatz, W. Witthuhn; Nuclear Phys. **A 239**, 253 (1975).
[28] D. v. Harrach, H. J. Specht, Nuclear Instr. Meth. **164**, 477 (1979).
[29] H. Bartsch, K. Huber, U. Kneissl, H. Sattler, Nuclear Instr. Meth. **121**, 185 (1974).
[30] A. S. Penfold, J. E. Leiss, Phys. Rev. **114**, 1332 (1959).
[31] E. Bramanis, T. K. Deague, R. S. Hicks, R. J. Hughes, E. G. Muirhead, R. H. Sambell, R. J. J. Stewart, Nuclear Instr. Meth. **100**, 59 (1972).
[32] B. C. Cook, Nuclear Instr. Meth. **24**, 256 (1963).
[33] V. E. Zhuchko, Yu. B. Ostapenko, A. S. Soldatov, Yu. M. Tsipenyuk, Nuclear Instr. Meth. **136**, 373 (1976).
[34] L. J. Schiff, Phys. Rev. **83**, 252 (1951).
[35] R. Vandenbosch, Phys. Rev. **C 5**, 1428 (1972).
[36] J. T. Caldwell, E. J. Dowdy, B. L. Berman, R. A. Alvarez, P. Meyer, Phys. Rev. **C 21**, 1215 (1980).
[37] J. M. Blatt, V. F. Weisskopf, Theor. Nuclear Physics (J. Wiley, New York 1952).
[38] A. Veyssière, H. Beil, R. Bergère, P. Carlos, A. Leprêtre, K. Kernbath, Nuclear Phys. **A 199**, 45 (1973).
[39] L. Katz, A. P. Baerg, F. Brown; Proceedings of the Second United Nations Conference on the Peaceful Uses of Atomic Energy 15, Geneva 1958.
[40] G. V. Solodukhov, Lecture Notes in Physics **62**, 216 (1977), Photonuclear Reactions II; G. M. Gurevich, L. E. Lazareva, V. M. Mazur, G. V. Solodukhov, B. A. Tulupov, Nuclear Phys. **A 273**, 326 (1976).
[41] I. S. Koretskaya, V. L. Kuznetsov, L. E. Lazareva, V. G. Nedorezov, N. V. Nikitina, Soviet J. Nuclear Phys. **30**, 472 (1979).
[42] J. Drexler, Thesis, Giessen, in preparation.

[43] Yu. P. Gangrsky, B. N. Markov, Yu. N. Tsypenyuk, Phys. Letters **32 B**, 182 (1970); Yu. P. Gangrsky, B. N. Markov, Yu. M. Tsypenyuk, Soviet J. Nuclear Phys. **11**, 30 (1970).
[44] H. C. Britt, Atomic Data and Nuclear Data Tables **12**, 407 (1973).
[45] T. Bar Noy, R. Moreh, Nuclear Phys. **A 229**, 417 (1974).
[46] J. Borggreen, J. Hattula, E. Kashy, V. Maarbjerg, Nuclear Phys. **A 218**, 621 (1974).
[47] P. D. Goldstone, F. Hopkins, R. E. Malmin, P. Paul, Phys. Rev. **C 18**, 1706 (1978).
[48] H. C. Britt, M. Bolsterli, J. R. Nix, J. L. Norton; Phys. Rev. **C 7**, 801 (1973).
[49] B. B. Back, J. P. Bondorf, G. A. Otroschenko, J. Pedersen, B. Rasmussen, Nuclear Phys. **A 165**, 449 (1971).
[50] B. B. Back, H. C. Britt, O. Hansen, B. Leroux, J. D. Garrett, Phys. Rev. **C 10**, 1948 (1974).
[51] B. B. Back, O. Hansen, H. C. Britt, J. D. Garrett, Phys. Rev. **C 9**, 1924 (1974).
[52] J. H. E. Mattauch, W. Thiele, A. H. Wapstra, Nuclear Phys. **67**, 1 (1965); K. A. Keller, J. Lange, H. Münzel, in Landolt-Börnstein, gr. I, vol. **5 a**, 1 (1973).
[53] R. Vandenbosch, Phys. Letters **45 B**, 207 (1973).
[54] S. Bjørnholm, V. M. Strutinsky, Nuclear Phys. **A 136**, 1 (1969).
[55] K. L. Wolf, R. Vandenbosch, P. A. Russo, M. K. Mehta, C. R. Rudy, Phys. Rev. **C 1**, 2096 (1970).
[56] S. M. Polikanov, G. Sletten, Nuclear Phys. **A 151**, 656 (1970).
[57] K. L. Wolf, J. P. Unik, Phys. Letters **43 B**, 25 (1973).
[58] E. Migneco, G. Russo, R. de Leo, A. Pantaleo, Phys. Rev. **C 16**, 1919 (1977).
[59] V. Andersen, C. J. Christensen, J. Borggreen, Nuclear Phys. **A 269**, 338 (1976).
[60] D. K. Sood, N. Sarma, Nuclear Phys. **A 151**, 532 (1970).
[61] J. Pedersen, B. Rasmussen, Nuclear Phys. **A 178**, 449 (1972).
[62] D. Habs, M. Just, V. Metag, E. Mosler, B. Neumann, P. Paul, P. Singer, H. J. Specht, G. Ulfert, Max Planck Inst. Kernphysik, Jahresbericht (1975) 55; P. Singer, Thesis, Heidelberg, 1976, unpublished.
[63] H. F. Brinckmann, D. D. Clark, N. J. S. Hansen, J. Pedersen, Phys. Letters **43 B**, 386 (1973).
[64] R. G. Clark, L. E. Glendenin, W. L. Talbert, Physics and Chemistry of Fission (IAEA, Vienna, 1973), Vol. II, 221.
[65] W. John, F. W. Guy, J. J. Wesolowski, Phys. Rev. **C 2**, 1451 (1970).
[66] J. W. Grüter, K. Sistemich, P. Armbruster, J. Eidens, H. Lawin, Phys. Letters **33 B**, 474 (1970).
[67] K. Bangert, U. E. P. Berg, G. Junghans, R. Stock, K. Wienhard, H. Wolf, Nuclear Phys. **A 261**, 149 (1976).
[68] M. Just, U. Goerlach, D. Habs, V. Metag, H. J. Specht, Proceed. Internat. Symp. Phys. Chem. Fission, Jülich 1979, Vol. I, p. 71, IAEA Publication, Paper SM/241, A 4.
[69] P. Limkilde, G. Sletten, Nuclear Phys. **A 199**, 504 (1973).
[70] P. A. Russo, R. Vandenbosch, M. Mehta, J. R. Tesmer, H. L. Wolf, Phys. Rev. **C 3**, 1595 (1971).
[71] H. Backe, L. Richter, D. Habs, V. Metag, J. Pedersen, P. Singer, H. J. Specht, Phys. Rev. Letters **42**, 470 (1979).
[72] H. J. Specht, E. Konecny, J. Weber, C. Kozhuharov, Internat. Symp. Phys. Chem. Fission, Rochester 283, (1973), IAEA-SM-174/19.
[73] R. Kalish, B. Herskind, J. Pedersen, D. Shackleton, L. Strabo, Phys. Rev. Letters **32**, 1009, (1974).
[74] D. Habs, priv. communication; Jahresbericht 1977, p. 46, Max Planck Inst. Kernphysik, Heidelberg.
[75] J. R. Huizenga, R. Vandenbosch, Phys. Rev. **120**, 1305 (1960).
[76] R. Vandenbosch, J. R. Huizenga, Phys. Rev. **120**, 1313 (1960).
[77] E. H. Auerbach, F. G. J. Perey, Report BNL 765 (Z-286), (1962).
[78] H. Bartsch, K. Huber, U. Kneissl, H. Krieger, Nuclear Phys. **A 256**, 243 (1976).
[79] H. Bartsch, W. Günther, K. Huber, U. Kneissl, H. Krieger, Z. Phys. **A 285**, 71 (1978).
[80] R. Vandenbosch, P. A. Russo, G. Sletten, M. Mehta, Phys. Rev. **C 8**, 1080 (1973).

[81] V. Metag, private communication, Jahresbericht 1979, p. 53, Max Planck Inst. Kernphysik, Heidelberg.
[82] H. H. Duhm, (Hamburg) private communication, 1981.
[83] N. L. Lark, G. Sletten, J. Pedersen, S. Bjørnholm, Nuclear Phys. **D 139**, 481 (1969).
[84] I. Mamamoto, W. Ogle, Nuclear Phys. **A 240**, 54 (1975).
[85] H. C. Pauli, Phys. Reports **7 C**, 35 (1973).
[86] U. Mosel, H. W. Schmitt, Phys. Letters **37 B**, 335 (1971).
[87] P. Möller, J. R. Nix, Nuclear Phys. **A 229**, 269 (1973).
[88] M. Bolsterli, E. O. Fiset, J. R. Nix, J. L. Norton, Phys. Rev. **C 5**, 1050 (1972).
[89] J. Libert, M. Meyer, P. Quentin, Phys. Letters **95 B**, 175 (1980).

(Received: 23. 6. 81)

## INHALT



---

Die „FORTSCHRITTE DER PHYSIK" dienen der Publikation zusammenfassender Berichte über aktuelle Teilgebiete der Physik. Neben Originalberichten in deutscher oder englischer Sprache werden deutsche Übersetzungen wichtiger fremdsprachiger Berichte abgedruckt.

Aus dem Inhalt der folgenden Hefte:

S. KAPTANOGLU, N. K. PAK: Technicolor, Extended-Technicolor and Tumbling

I. H. DURU, H. KLEINERT: Quantum Mechanics of H-Atom via Path Integrals

A. ACTOR: Nonexistence of Multiple-Instanton Solution of the Massless 4D$\phi^4$ Theory

H. KLEINERT: Quasiclassical Approach to Collective Nuclear Phenomena

H. KRESSE: Dielectric Behaviour of Liquid Crystals

S. M. BILENKY, S. I. BILENKAYA, G. MOTZ: Neutral Current and $P$-Odd Effects in Processes with Participation of Leptons

I. BENDER, H. C. DOSCH: $N$-$N$ Potentials in QCD-motivated Quark Models

Manuskripte werden nach Vereinbarung mit dem Autor von der Chefredaktion entgegengenommen.

Bei Fragen zur Produktsicherheit wenden Sie sich bitte an:
If you have any questions regarding product safety,
please contact:

Walter de Gruyter GmbH
Genthiner Straße 13
10785 Berlin
productsafety@degruyterbrill.com